THE EVOLUTIONARY JOURNEY

A Personal Guide to a Positive Future

BARBARA MARX HUBBARD

Evolutionary Press
of the
Institute for the Study of Conscious Evolution

ISBN 0-943408-01-6

Typeset and Produced by
Word Masters™
820 NE 45th St.
Seattle, WA 98105

Cover:
Original Painting
by
Scott Toms

**Published by
Evolutionary Press
2418 Clement Street
San Francisco, CA 94121**

Printed in U.S.A.

CONTENTS

ACKNOWLEDGEMENTS

I acknowledge my partner John Whiteside's vision and encouragement. He asked me to write down every idea I wished to communicate and from those ideas he created "The Theatre for the Future — Previews of Coming Attractions," a multimedia theatrical which I narrated live. This book is taken from *The Evolutionary Journal* which I put together from the Theatre for people who had seen it and who participated in ACT III, a multimedia seminar also designed by John.

I also acknowledge the original paintings done by Los Angeles artist Scott Toms and by Earl Hubbard which were first featured in the Theatre and are reproduced in this book.

I am grateful to Dr. Roderic Gorney for permission to quote from *The Human Agenda* and to R. Buckminster Fuller for permission to quote from *It Came Not To Pass But To Stay.*

February 14, 1982
Valentine's Day
Washington, D.C.

Dear Friend of the Future,

Welcome to *The Evolutionary Journey.* Welcome to the most exciting adventure since the dawn of self-consciousness.

If you feel a deep desire to do more, to be more; if you intuit a "memory of the future;" if you are attracted to the unknown and wish to participate in your own evolution and the evolution of the world, this little book will appeal to you. It is an early guide to "conscious evolution." It maps the process of Creation from the origin of the Universe to our potential future. Through special ideas and key readings, *The Evolutionary Journey* will help you find your true vocation, your own next step forward.

Yours for a Positive Future,

Barbara Marx Hubbard

Barbara Marx Hubbard

FOREWORD

This is a book about potential, our — the human species — collective potential to transform our world from one of distress to one of joy and excellence. *The Evolutionary Journey* appears at a time when it is deeply needed. There is today an abundance of confusion and apprehension about the future. Our questions are deep and disturbing. How will we solve the seemingly insurmountable social and ecological problems we have created? Will we make it? Is there any hope?

We all wish for hope, yet we need an inspirational context, a supportive foundation upon which to base our hope. Barbara Marx Hubbard offers this.

For some twenty years Barbara has been recognized as one of the leading inspirational thinkers in the field of Futurism. At the same time, it is only honest to say that her views often have been regarded as far-fetched, overoptimistic and ungrounded in the "realities of our time." So be it! Such has always been the case with pioneers who challenge the prevailing ways of seeing the world.

Over the years Barbara has lectured widely. However, her ideas have only reached the broader public in her earlier work, *The Hunger of Eve* (Stackpole, 1977), which is largely autobiographical. Thus there has been a gap in the public presentation of her vision. *The Evolutionary Jour-*

ney now fills that gap, and does it well. Herein is an excellent summary of her broadest and highest inspiration; herein is a view of reality that touches the heart, opens the mind and inspires hope.

There are three specific qualities that make this book a landmark for those interested in the future: 1) a broad evolutionary perspective, 2) a profound spiritual vision and 3) a hopeful, positive orientation.

1) For the most part we of the Western World have been taught to regard ourselves as products of a few thousand years of social/political history. While this perspective is not inaccurate, it is clearly short-term and therefore limiting in its implications. In contrast, the evolutionary perspective views us as emergent aspects of some 15 billion years of evolutionary development. In this light we come to regard ourselves as magnificent, yet unfinished, expressions of the guiding intelligence behind evolution. We see ourselves, in the midst of change, reaching forward toward unknown potentials. We see that we are not only in process, but that we are the process itself; we are the leading edge of evolutionary transformation. Hence we now recognize that we can play a significant role in determining a future that is as awe inspiring, if not more so, than our evolutionary past has been.

2) *The Evolutionary Journey* springs from a deep understanding of the Universe as a divinely guided emergence. In direct lineage from Teilhard de Chardin, Barbara imbues her words with a poetic light that reflects the inner light in each of us —we are, thereby, "enlightened."

3) Barbara's unwavering positive orientation is of special importance. Much evidence has recently emerged in both physics and parapsychology that indicates a direct relationship between

the feelings, thoughts and beliefs we hold and the reality we experience. The power of the human mind, however, goes well beyond bending forks.

It is my belief that the quality of the collective human consciousness can directly influence, for better or worse, widescale planetary phenomena, both social (crime, war, etc.) and geophysical (weather, earthquakes, etc.) Bearing this in mind we can now appreciate the value of holding a strong, clear and positive orientation towards ourselves, our world and the Universe. The picture that *The Evolutionary Journey* portrays is one of hope and inspiration — necessary ingredients if we are to consciously evolve toward a new age of peace and transcendence.

The Evolutionary Journey has been envisioned as a companion volume to my book, *Conscious Evolution* (Evolutionary Press, 1982). Both are introductions to the growing field of Conscious Evolution. *The Evolutionary Journey*, I believe, will prove to be one of the key texts in this integrative field. In fact the three qualities discussed above — evolutionary perspective, spiritual depth and positive orientation — can provide guidelines for others who wish to contribute to the field.

Throughout the text, key ideas are highlighted, reframed and restated a number of times. The aim is to teach — to open our eyes to a broader perspective that will nurture our growing understanding of ourselves as potential agents of conscious evolution. Further, since Barbara's vision is broad and general, it does not include either theoretical or practical guidelines for work in any specific field. This work is left to others who are willing and able to focus in a particular area — economics, education, environment, space exploration, etc. — within the broad, encompassing context of conscious evolution.

This is perhaps the central contribution that the field of Conscious Evolution will make to the future — to offer a unifying concept within which more specific studies and disciplines can relate to one-another and find their meaning.

I am delighted that Barbara Marx Hubbard has offered us this inspirational text which helps to lay the foundation for the enterprise of conscious evolution. *The Evolutionary Journey* is one small, but significant step toward a positive future.

Barry McWaters
Mill Valley, California
April, 1982

PART I

SETTING FORTH

We Hold These Truths

. . . There is an evolutionary process
of Creation
which has been
and is now
progressing toward
greater
consciousness,
freedom and
purposefulness.

. . . The Creation is continuing
and we are
a part of it.
We are the womb
of the future,
the co-authors of
Creation.
Our destiny is
to share with
the Creative Intention,
to take more responsibility
for this Intention.
As our awareness of it grows,
we are becoming
co-creators.

. . . Our lifetime marks
the birth of Universal Humanity.
We are the first generation
to be aware of
ourselves as one being.
We are the first
to accept responsibility for
the future of the whole.

. . . We are one body
born into the Universe
discovering greater awareness
of our Creative Intention.
Earth bound history is over,
universal history is begun.
We are born into a Universe
of immeasurable possibilities.

. . . Our image
of the future
affects our actions.
As we see ourselves
so we act;
as we act,
so we become.

. . . Our future is
a contingency,
not an inevitability.
It depends upon
our acts.

Chapter 1

DISCOVERING PURPOSE

Deep in our consciousness is the need for purpose. We yearn for significance — over and beyond all personal goals. It is for relatedness to a higher order that the flame of expectancy burns.

Great ages in the past had a common transcendent vision which attracted the energies of the majority, even if unconsciously. In ancient India, Egypt, Greece, and in all great cultures at their height, the flame in each person joined multitudes of others to rise beyond the scope of the individual self.

But in our age, past visions of meaning have

broken down. Our generation was born during a great pause, a dark night of the soul of the Western World.

We all experience this lack of common purpose, yet each in a different way. I felt it, even as a child, as a fundamental need for meaning that I could not live without. In 1948, when I was eighteen, I wrote in my journal:

> "It's Christmas but I feel none of the mystery, the peace, or the warmth. All the beautiful feelings that come to one on Christ's birthday shun me. Instead I'm tortured with doubt, fears and unhappiness. There's a constant pull in the middle of my stomach. I'm torturing myself to death. The cause is evident. In my own eyes I've achieved nothing, yet those same eyes have visions of glory untold. I must either lower my ideals or achieve them. I'm like a magnet feeling the attracting force of another magnet, yet held apart."

Ever since, my unfolding discovery of the evolutionary vision has been guided by this magnetic attraction.

MY STORY

I was born in New York City in 1929 in a good and kindly family — agnostic, secular and generous. I had no external "problems" to divert my attention. I suppose I should have considered myself fortunate and felt content. And yet, my earliest memories are of a deep hunger for something more, for a purpose beyond my daily life.

This nameless desire, pulling at my solar plexus like a magnet, was making me miserable. What would happen, I wondered, if most people had "enough" material sufficiency? I could already sense the impending gloom of the whole human species arriving at the goal of material comfort . . . and then feeling the same longing as I. They too would ask the same questions, feel the same need.

I lived through the Second World War, relatively unaware of the horror. Then in 1945 the atom bomb exploded. It triggered a deep question in my mind: What is the purpose of Western Civilization? What is the meaning of science, technology, industry? Is it leading to destruction? Or even worse, is it leading to meaningless affluence? Neither seemed right. For inside myself there was still this magnetic attraction to the future that made me feel sure that something new and wonderful was about to happen! But what could it be?

I started on a search. What is the desirable direction of the future?

At the time, in the late forties and early fifties, the modern arts were portraying the end of an age. From the Renaissance masterwork, "David" by Michelangelo, through Manet, Monet, Pissaro, Picasso, Jackson Pollock, we witness the disintegration of a self-image. T. S. Elliot and Jean Paul Sartre heralded the mournful meaninglessness of modern aspirations for progress. Hope was ridiculed; aspiration was assassinated. We had come to a dead end.

I read like a child starving for life. First, the world philosophies. What do they say about the future? Very little indeed. Either they are reactionary looking backward toward a golden age; or stoical, accepting everything as is; or repetitive, expecting nothing new under the sun; or existential, asserting that there is no intrinsic meaning except what you make up; or absurdist/nihilistic, basing their expectations on the inevitable increase of entropy and looking forward to the eventual heat death of the Universe. Each of these philosophies seemed equally bleak in its outlook, and since none of them resonated with my intuitive hope, I was still dissatisfied.

I turned to the world religions, looking for one thing — an image of the future commensurate with our power and aspirations. I found them much more interesting! They all "predicted" transformation. Especially the Judeo-Christian which suggested a time would come in history when, as St. Paul said, "You shall not all sleep . . . you shall all be changed . . . in the twinkling of an eye, at the last trump, and the trumpet shall sound, this mortal flesh will become immortal and death shall have no dominion . . . the sufferings of the present cannot be compared with the glory which shall be revealed in us . . ."

There! The magnetic pull was for "that." But what is "that," and how do you get from Scarsdale, New York, to glory untold?

I tried to make the mystical leap of faith from here to "there," but my rational mind would not go. I failed, falling into the metaphysical abyss between secular reality and spiritual faith.

In desperation I joined the Episcopal Church in Scarsdale — a mecca of comfort. The minister preached we were helpless and guilty. I wanted to rise up and say, "It's not true! We must be great! God would not have created us helpless. He created us in his image." But I did not dare and, besides, when I prayed I heard no voices, saw no signs. Only silence. I despaired and left the Church.

To every young man who ventured to take me out, I asked the question, "What is your purpose?" "Why do you think we're here on this planet?" My dating life was slow . . . to say the least.

I tried academia — Bryn Mawr College. My heart sank the first day. The courses were all in separate boxes: History, Mathematics, French

Literature. No course on purpose. No relationship between ideas.

People said, "When you marry this will all pass. Motherhood is obviously the answer." I married and had five children — a thorough investigation — yet the need for purpose deepened. I loved each child as myself, and needed to know their meaning as well as my own.

By the time I was thirty I thought I was neurotic. I had done everything I could think of to be purposeful. Love. Children. Husband. I had clean air and water, good food. I tried to make people "happy." And I was dying. But of what? Of meaninglessness.

Then I made three discoveries.

First, Abraham H. Maslow's *Toward a Psychology of Being*. It saved my life. He said that once your deficiency needs are relatively easily met, your "growth needs" for a richer sense of beingness, for chosen work of intrinsic value, become imperative — as necessary as food to the hungry. If you don't grow at that stage, you get psychologically sick.

I then understood that my problem was unused growth potential. I simply had not known what to work for. I was underdeveloped, not sick! Alienation was a divine discontent driving me toward a next level of growth. I determined to learn how to be "normal" and joyful in the 20th Century.

Maslow also pointed out that all self-actualizing people feel related to a transcendent or transpersonal order beyond self-fulfillment. This too felt right to me. It explained, quite simply, my urgent reaching out toward the source of meaning, toward the divine.

The second life-saver was Teilhard de Chardin's *Phenomenon of Man*. An epiphany! There is a continuing, evolving pattern in the process of nature which leads to greater whole systems, higher consciousness and freedom — and it's going on NOW! It's unfinished. The world is evolving, not just the individual. Not only do I have unused growth potential — so does the world — so does our species — so does the Universe! Something new is coming. The magnetic attraction was right. I could trust my intuition.

I was getting closer to the answer to the question of purpose.

Then in 1962 John Glenn penetrated our blue biosphere into outer space. I had a sense of our species at the moment of physical transformation. Not only is our consciousness expanding as the mystics have proclaimed for ages . . . so also, our bodies are transcending Earth-boundedness. When I saw the rocket rise the words FREEDOM! BIRTH! exploded into my mind. We could become a universal species! That was the purpose of our power, our technologies, our sciences. They provided us with the skills to carry us beyond the womb of our Mother Earth into a universe in which there may be billions of life-systems comparable to our own. The Western World has been developing the technologies of transcendence commensurate with the visions of transcendence.

It was like being alive when the first fish flopped out on dry land . . . a critical evolutionary event that surely went unnoticed by that adventurous fish and his friends. Similarly we, the human family, have not yet understood the significance of present events. We are still lacking a perspective broad enough to see what's happening. However, for me, something began to change radically as I opened to a bigger vision.

In one of those flashes of insight and expanded reality that can change one's life, I experienced ourselves as one planetary body being "born into the Universe." In my mind's eye I was in outer space with that rocket, witnessing myself and all others as conscious cells in the body of Earth. Our whole body was struggling to coordinate. We were gasping for breath as pollution poisoned our biosphere. We were feeling the pain of oneness as the mass media flashed "news" of hunger, accident, and war. And as we reached together into outer space for new life, we awakened to our oneness, we connected and we experienced total empathy, total joy. The whole planet smiled — a planetary smile!

The flame of expectancy rose to a burning intensity, motivating me to seek an expanded vocation, an "evolutionary connection" between my need for personal growth and the needs of the world in transition. My second life began — a chosen life. I felt free within myself not by separating myself from the formative processes of the Universe, but by connecting my own growth urge to the force that is evolving the Universe. I had discovered a purpose related to the evolution of the world.

I worked with a powerful joy to try to understand the Story of Creation at the dawn of the Universal Age. I read constantly, piecing together cosmology, geology, biology, psychology, current events, futuristics, religious visions. I sought out people throughout the world working on aspects of our next step forward, feeling the attraction which Teilhard describes. All my relationships changed. I became a growing woman, a deeper mother, able to relate to my children as co-evolvers in a transforming world. In my forties I felt at the beginning of life — almost a second adolescence.

I became convinced that we stand upon the threshold of the greatest age of human history. Anyone who can see the opportunities will emancipate their own potential through involvement in the world. Personal and planetary growth are not merely parallel, they are one. The individual cannot flourish in a dying world, and the world cannot evolve if its members are alienated and in despair.

The key to the next step is awareness of what we can become, as individuals and as a whole species. This is the purpose of *The Evolutionary Journey* — to see together what we can become based on what we have been. Together we shall map the story of our creation and travel into the unknown.

Chapter 2

MAPPING THE JOURNEY

We live at the most marvelous moment in human history. Everyone now alive is involved in the greatest upheaval since Humanity emerged out of the animal world.

We are at the dawn of "conscious evolution," when the creature-human first becomes aware of the processes of Creation and begins to participate deliberately in the design of our world.

We are the first generation to awaken to the awesome fact that we are affecting the future by our every act, from the number of children we have and the kind of food we eat to the creation of new life forms and new worlds in space.

We will either stumble blindly toward self-destruction through misuse of our new powers, or we will move consciously toward a new order of the ages. We will begin the great evolutionary tasks: the restoration of Earth, the freeing of people from want, the development of the vast untapped potential of our bodyminds and the exploration of the unlimited frontiers of outer space.

Different people are drawn to different vital functions, often not realizing they are interconnected. Environmentalists preserving the biosphere may not recognize that they are related to space scientists extending the biosphere. People exploring the inner space of spiritual growth may not see their relationship with political leaders agonizing over the allocation of resources. Those concerned with freedom may not identify with others working toward greater global unity, etc. Nonetheless, all are related in one complex planetary act. A powerful movement of action for life is spreading throughout the world. Untold individuals are building a "cathedral of action," an edifice of effort to transcend the limitations of the present and to create a desirable future for humankind. Each embodies a gift, a shining act of excellence which is incorporated into the cathedral.

There are three invisible pillars to this cathedral: Freedom — all acts to emancipate us from deficiencies in order to do our unique best; Union — all efforts to unify the world in freedom; Transcendence — all breakthroughs to relate us consciously with the universal processes of which we are a creative part.

Increasing numbers of people, experiencing the transformation with joy, rather than fear, are gathering together to ask, to listen, to learn how to take our next step into a future of immeasureable possibilities.

The choices are unprecedented. The situation is new. And we are the actors in this cosmic drama. What is our story? Where is our script? How shall we know what to do?

MAPPING THE PATH

The Evolutionary Journey provides a new context to discover answers to these questions. It is a preview of the actual journey. As we embark, we will fuse the eyes of science and intuition to discover the story of our creation. We will cast the searchlight of our mind deep into the distant past to witness the origins of the Universe, the formation of our solar system and Earth, the appearance of single-celled life and the origin of the animal world, leading to the dawn of Humanity. We will observe ourselves, now, undergoing another evolutionary turn of the spiral, at the beginning of the Universal age.

From this perspective we will see our future as a natural continuation of fifteen billion years of transformation. We will seek guidelines in the recurring patterns of Creation in order to apply them to our choices now.

We will piece together a positive vision of the future based on the reality of our potentials — catching a glimpse of glory — ourselves in the future, a universal species, beyond self-centered consciousness, beyond planet-boundedness, beyond unchosen work, beyond disease, hunger and war, about to touch the "tree of life" as we enter the universal phase of our history. We are a young, uncertain, unfinished species, with the genius not only to survive but to prevail, to grow, to change, and to become at last fully human, partners with the processes of creation, ever-evolving in a Universe without end.

Then, in this breathtaking context, with a new vision of the future commensurate with our emerging capacities, we will look at the questions: What can we do now? What are our personal and social goals? How can we expand our perspective and become agents of conscious evolution, linking with others, wherever they may be, in a magnificent world-wide network of shared attraction to that which we all may become?

WHAT IS THIS FLAME?

What is this flame of expectancy, this growth impulse, this frustration which urges us to break through the given pattern of our lives, pulled as though by an irresistible magnet toward each other and toward a desirable yet undefined state of being?

The flame feels like the binding, magnetic force that runs through the process of evolution, the force that drew atoms to atoms, molecules to molecules, cells to cells and now people to people in ever more complex whole systems. It feels like love. Not only personal love for each other, one by one; nor only altruistic love for others; but also, and more so, a transpersonal love, which senses the fulfillment of one's self through deep involvement in the evolution of the world, in connection with a Universal Force.

The flame feels like the attractive and attracted energy that moves the process — the intention of Creation. It represents our contact with the Logos, the Designing Intelligence, the Creative Intention — God.

It was ignited at the beginning of Creation. It became self-conscious at the dawn of Humanity.

It is now rising higher, kindled by the maturation of science as it probes the mysteries of the Universe from without, and by the deepening intuition of millions attuning to reality from within — and finally by the dangers pressing upon our species as a whole.

It will fuse with our own awareness as we learn "conscious evolution" and become partners with the God-force, co-creators of our futures.

Everyone is a part of this planetary drama. Each of us is like a cell in a planetary body which is undergoing its "birth" from its terrestrial, self-centered phase to its universal, whole-centered phase. We can no more escape this condition than a cell in the body of a baby in the birth canal can opt out of the game. Whatever happens, happens to us all.

Perhaps the key difference among us is our response to this evolutionary change. Some react in pain and shock, withdrawing, resisting, entrenching. Others are neutral, waiting to see how it will turn out, ready to accept what the fates may bring. But there is a third group, widely scattered throughout the world, from all backgrounds, races and nations, who are now sensing a powerful, personal energy of creativity coursing through their lives.

The flame of expectation is igniting us, animating and attracting us to move beyond our existing life patterns . . . to do more, to be more, to know more, and above all to connect with each other, to discover what we can do together that none of us can do isolated and alone.

Do these words seem truthful to you? Have you experienced yourself as a living part of the Creation, at one with all nature?

Do you sense a Designing Intelligence at work in the Universe? Do you feel a "memory of the future" as though at the depth of your being you already know more than you can yet fully understand?

Do you desire to participate in the evolution of yourself and the world? Do you feel close to the stars, drawn toward the Universe beyond Planet Earth?

Do you feel a love too great to be contained by personal relationships? Do you long to connect and commune with those who share your attraction?

If so, you are a member of the growing future-oriented family of Humanity, beginning now to meet other members of your family. When you meet another who shares this sense of hope and excitement, you know each other immediately. The connections are now "thickening" as more and more of us converge, finding each other and discovering our common tasks, our chosen vocations, our natural next steps in the evolutionary journey.

If a picture of Earth from a satellite could view, not the physical but the psychological weather, we would see an intense build-up of flaming pressure in clusters and networks — like small bush fires — heating up with psychological tension and sensitivity, ready to burst beyond the walls of self-centered consciousness to fuse in concerted action with like-minded people. This pent-up energy is causing mental disturbances, breakdowns, ecstasies and visions. We are building up to a storm of creativity. The pressure awaits unknown triggers that will release the tension and spread the flame.

From space we would see all this as a brilliant flaming light spreading throughout the globe, awakening those in whom the flame is dormant, until most people upon the planet are aware that they are working together as one body of almost infinite diversity, infused with the energy of Creation.

PART II

THE JOURNEY

The Theatre of the Future

We are now embarking on the evolutionary journey of our species. Deep in the distant past we will seek out our roots at the origin of Creation. We will trace our history through billions of years of cosmic evolution to the present, ourselves on Planet Earth, awakening for the first time to the awareness that we are participants in the Creation.

As you read, visualize a theatre for the future. The narrator stands before a large screen upon which we can imagine the images of the past and the future. As Pachelbel's glorious "Canon in D" begins to play you see an artistic rendition of an evolutionary spiral crawling across the galaxies. Then you see your little home Planet Earth . . . and the great historical events: the origin of life, the dawn of Humanity, and our birth into the Universe. You see the "synergizing Earth" linking up with points of light until the whole planet glows. And you see a visualization of a "haloed Earth," as we overcome our problems and emerge as one planetary body, born into this Universe, seeking greater contact with life.

The text in Part 2 was written to trigger your memory of the future by awakening the memory of the past. Please read it aloud to yourself, your family, your friends. Allow your mind to freely visualize the magnificence of the journey. Please permit yourself to experience the profundity and joy of evolution unfolding.

The ideas are brief and punctuated. You can pause at any point and close your eyes for a few moments. Images from your own inner wisdom will appear to fulfill and complement the text.

Now, turn down the daylight and step into the Universe. Think of each idea as a separate thought in the great mosaic of cosmic history. Visualize a stellar stained glass window stretching across the skies, our media cathedral, our theatre of the future. Witness our journey of billions of years gone by and billions of years to come . . . a preview of coming attractions.

Chapter 3

THE PAST
Fifteen Billion Years of History

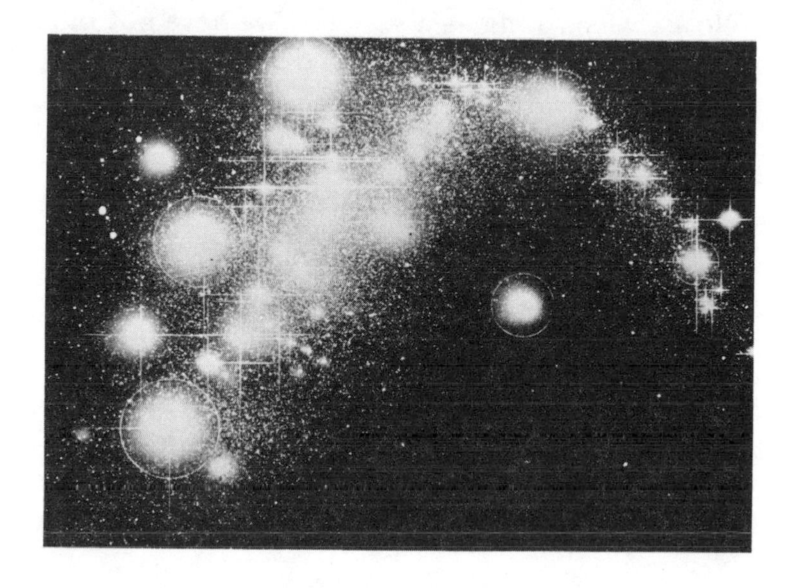

Imagine the Universe! Billions and billions of galaxies, multitudes of solar systems, some that may have life comparable to our own.

Narrow your vision to a local event — our galaxy, the Milky Way — and now a very tiny, almost invisible speck of dust — us — Earth. We are so small that when an astronaut looked at us from outer space, he hid our world behind his thumb.

Step outside our planetary system, project yourself into the future and imagine that we are already a more mature species — imagine what it might be like when we grow up, stop fighting and start using our capacities harmoniously.

Just as each individual has a higher, wiser self towards which we are growing, so our species as a whole has a greater species potential — that which we can become as a whole society of maturing individuals.

• • •

It is impossible to judge the human race yet. We are very young as a species. We have nothing to compare ourselves with. We have never seen another planetary civilization go through this transformation. We are like a newborn baby who has never seen another. We can look upon ourselves with compassion and reserve judgment.

In this story we are living in the future, looking at our own past as a movie of Creation, a "photogenesis," accelerated in time. Imagine the rapid film sequence of the growth of a plant from the bulb, through the green shoots, to the unexpected burst of the flower. If you were a bulb, and had never seen a flower — could you have predicted the radiance of the bloom? Now we are a planet bursting into our next phase. Can we predict what we are to become?

• • •

Our story of Creation covers billions of years. It is an unfinished story. The Creation is not over! Most stories of Creation start and end — but ours obviously is continuing. We are part of the Creation. We are affecting the future by our every act, thought and desire. This is a miracle play in which we are the actors.

The story of Creation is our story. We already know it, because it has already happened to us. We are evidence of fifteen billion years of success. The atoms in our body were formed at the beginning of the Universe. The iron in our body was formed on the early Earth. The cells in our bodies have the memory of the origin of life. In our brain the reptilian and mammalian experiences are still monitoring and influencing our unconscious life.

If we trigger the memory of the process of our Creation, we will remember ourselves. Within each of us is the experience of the past, and perhaps the pre-figuration of the next phases of growth.

To fulfill Socrates' great commandment, "Know Thyself," we must trigger the memory of the entire process of our Creation. For our self's roots are in the Infinite.

To know where we have been and where we are going we need a new cartography. In the great Age of Exploration, the maps were inaccurate. Columbus started our for the Indies and bumped into the American continent! But with each voyage, with every pioneer, the maps improved. Now we can go anywhere on Earth with the knowledge that someone has been there to help guide our way.

The new map is not geographical, but a map of Humanity as a process evolving in space/time — an evolutionary spiral.

At the beginning of the Universal Age, the map of the process of our Creation will be inaccurate. There can be no dogma. The knowledge is uncertain. But we are surrounded by information which is flooding in from all areas of research and experience. We are pioneering in awareness, unraveling the mystery of our past for the first time. All information is subject to correction. The evolutionary spiral is a map upon which to develop the picture of our cosmic past, present and future.

• • •

Our spiral map starts with a glow, a flame of ineffable light — the Logos, the Creative Intelligence — God. We are witnessing the outward and visible signs of an inward process. This is an esoteric mystery story. We can see the physical manifestations, but the intention, the consciousness is unseen. The real star of this mystery play is invisible.

It all began fifteen to twenty billion years ago with the creation of the material Universe.

Four to five billion years ago, our solar system and Earth were born.

Three and a half billion years ago, single-celled life emerged.

Eight hundred million years ago, multi-cellular life developed: plants, animals, the biosphere.

Then, two to three million years ago, early Humanity.

And now, it's us! We are going round the loop. It's unknown terrain. We're mapping it now. We feel it in our own lives as the speeded-up experience of change — a leap from one condition to another new and different condition — a "quantum transformation" — as from molecule to cell, or animal to human. There is as yet no agreed-upon name for the next phase of Humanity.

• • •

Looking back on the fifteen billion years of our evolution, certain "lessons" can be learned. **Crises precede transformation**. Before every quantum change, "problems" emerge — limits to

growth, stagnation, unmanageable complexity, impending catastrophies, disintegration. From the perspective of the present, the crises look like mistakes, deadly errors in the system. But from the perspective after the quantum transformation, these problems are seen to be "evolutionary drivers," vital stimulants which trigger astounding "design innovations." The crises trigger inventions which bring out new potentials and capture them in stable form and produce absolute newness — like the cell from the molecules, the human brain from the animal brain, or some next possibility arising out of our own limits and catastrophies.

From the universal perspective we see problems as vital evolutionary drivers toward new forms of life, necessary to overcome the inertia of the status quo.

What were those drivers, limits, catastrophies, innovations in the past, and what are they now?

Another lesson the spiral reveals is that **the creation of newness is an evolutionary fact.**

At one point there was no Earth, then there was Earth.

At one point there was no life, then there was life.

At one point there was no Humanity, then we appeared.

Whatever emerges from our quantum transformation will be new. It is implicit in the process that something new will emerge. Incremental changes do not lead up to more of the same.

• • •

The evolutionary spiral starts at the beginning of Creation. We do not know the nature of the conception — the ultimate mystery. When we put telescopes in space we will be able to see the light from the first explosion, real moving pictures of our physical creation. We on Earth are looking up through a dense atmosphere, like fish peering through the waters at the world. Now we see through a glass darkly. Soon we may see face to face. We have intuited the story of our Creation with the inner eyes of revelation. Now we add the young eyes of science, the new sensory system of the body of humankind. We sense from within and we see from without. The recovery of our past is a miracle of the modern age. Only now could this version of Creation be told.

The divine and the secular visions begin to fuse.

• • •

We imagine a dialogue between astronaut Frank Borman circling the moon on Christmas of 1968 and Professor Roderic Gorney writing in *The Human Agenda*:

Borman: In the beginning, God created the heaven and the earth. The Earth without form and void . . .

Professor: A thermonuclear blast vaporized the original ball of matter. Firey gasses spewed in all directions at the temperature of atomic fusion. The Universe was born. Time began. Within the first hour were formed the myriad miniature universes of the atomic elements. After a quarter million years of expansion and cooling the tugging cords of mutual magnetic attraction at last overcame the declining energy that had maintained the particles' even dispersion.

In some places the now chilled particles drew together turbulent clouds of gas and dust . . . cooling and condensing, solidifying and impacting, they fell together into larger and heavier masses. Evolving through billions of light years of black space, the clouds clustered into billions of isolated spinning galaxies, including our Milky Way . . .

Borman: And darkness was upon the face of the deep . . .

Professor: Clouds crushed inward with sufficient gravitational force to re-ignite nuclear combustion. The Milky Way began forming more than 150 billion stars . . .

Borman: And God said, let there be light . . .

Professor: Thereafter the Universe was lighted and warmed again near each radiant sphere . . .

Borman: And there was light. And God saw the light, that it was good . . .

Professor: In this fashion, five billion years ago, the center of a whirling disk-shaped cloud at the edge of the Milky Way contracted into an incandescent hub . . . a medium sized star we call the sun . . . small clusters circling in the web of its gravity condensed into the planets. One became our Earth . . . mother of life . . . midwife to the human agenda.

• • •

Our mystery play begins to unfold — the first and second loops on the spiral: the formation of the Universe, our solar system and Earth.

The "Big Bang" theory, the earliest event of which we have a record, suggests a Prime Cause, a Creator — the first evolutionary driver. Science in its maturity has confronted the ultimate mystery and now deepens its awareness beyond mechanistic explanations to seek out the non-

material Source, the Matrix, the Intelligence of that which it measures.

• • •

We know something of the evolutionary driver for our solar system and Earth. The cosmic gases suffered a heat loss and "adapted" by condensing and contracting into solids. The power of attraction was at work . . . leading to collapse, catastrophe to the status quo — and a design innovation: the SYNTHESIS OF ELEMENTS that make up the Earth and our bodies now. From hydrogen and helium atoms, collapsing in the crucible of burning stars, the minerals and metals in our bodies were formed four billion years ago. We are the living products of that design innovation. The complex process of geo-genesis is now being studied by cartographers of the new age.

• • •

Lessons emerge: the balance of nature is continuous, progressive change, with a recurrent pattern in the process: the creation of new whole forms out of the old through systhesis of separate parts. There is no point where the process remains static. The tendency in nature to form ever greater whole systems which are different from and greater than the sum of their parts, is intrinsic — or we would not be here. Trillions of cells organized into our complex bodies on a planet organized out of atoms and molecules, in a solar system circling a galaxy which is one of billions of galaxies in a Universe whose dimensions are still beyond our imagination . . . but is composed of wholes within wholes — ever evolving.

THE CREATION OF NEWNESS IS AN EVOLUTIONARY FACT

There was as yet no life in the silent seas of the early Earth. For untold millions of years amino acids, proteins, enzymes — the chemicals of life — formed in the waters of the world. The moon and tides filled and flushed the seashore pools. Lightning struck and volcanoes erupted. Large molecules developed, organized through the power of attraction that brought atom to atom.

Gradually, the molecules became over-extended, over-complex — perhaps like our societies today.

At some point the situation became "critical" — it could not continue as before. Evolutionary drivers were at work. Either disintegration or a higher level of integration would occur. (We can begin to identify with the experience of the entities that form us.)

• • •

Another design innovation appears — the GENETIC CODE. The DNA intelligence enters history as the first self-replicating molecule that

could divide and reproduce. Over billions of years these pre-life chemicals continued to grow in organization and complexity until finally, about three and a half billion years ago, the DNA intelligence "learned" to build the cell. Life! Sentient consciousness — a higher synthesis composed from the large molecules in the rich chemical soup of the seas. Somewhere, in the presence of an extra influence as yet unknown, the genetic code, the self-replicating capacity, took hold — the beginnings of the fabulous information technology now hidden in every cell carrying the coded blueprint to build a new unique body.

The origin of life is the next quantum transformation. We do not yet understand how it happened... the new cartography is still blurred.

Perhaps life, seeded on this planet from extraterrestrial origin, holdswithin it the pattern for the universal future — just as we, as a single fertilized cell in our mother's womb, held our own future curled up in the DNA. Is there a planetary building plan, a "cosmic DNA" guiding the evolution of a planetary system as there is of all biological systems? We will probably not know until we have other planetary systems to compare ourselves with.

We can imagine ourselves as the over-extended macro-molecule passively floating in the early seas, not expecting anything new — but feeling vaguely uncomfortable, sensing something "wrong" — something about to happen. We begin to sense our unused potential — when suddenly we are drawn by the RNA, the messenger of the DNA, into a new pattern, a larger whole — the cell. We are becoming protein — a different matter! "What's happening? I feel strange! I can see! Future shock! A New Age! I'm more than I was! I like it! I hate it! Something artificial is happening! Who am I now?..."

Imagine, after this link-up, falling out of the new unstable pattern. We remember the experience of being part of a larger whole, we remember the sense of cellular consciousness, and want to reconnect to feel "it" again.

Maybe it's like we feel today when we participate in a totally loving group, fuse with others, take on the power of the whole, and transcend our individual limitations.

Then, later, we fall out of the pattern and miss the joy of union. We feel the pain of separation and long to connect again.

Life appeared where there had been no life. Again newness. We learn to expect the unexpected — anticipate the new. Yet we learn the nature of the unexpected. There was a leap in awareness, purposefulness and freedom of action. Evolution is a consciousness-raising and freedom-raising experience. A cell has more awareness and freedom than a molecule.

The same pattern was at work — over-com-

plexity, higher synthesis, a new whole — the cell, a quantum leap in awareness.

More complex whole patterns bring out the potential of the separate parts.

• • •

Higher consciousness and more complex technological bodies evolve concurrently. They are the psychological and physical aspects of the same event. There is a *technology of creation* in every physical body. A continuum of technological capacity accompanies the rise in consciousness and freedom. For every advance in consciousness, a more complex technological body developed. Technology is not new. It's simply that human beings are new at it! We are the products of the most sophisticated technologies of Creation.

• • •

There was no sexual reproduction in single cells. Nor was death programmed into the individual. They were semi-immortal, our ancient single-celled ancestors. They divided to reproduce. It was a tremendous innovation for them.

Life became successful. The cells spread rapidly. They did not adapt to the environment. They are the environment and transformed it into themselves! A very important lesson. The organism acts on the environment as well as reacting to it. Life transforms and utilizes resources with an intrinsic purposefulness.

The cells absorbed the nutrients of the early Earth, prospering for a billion years. There was an apparent balance of nature, an apparent "steady state" economy. But something new began to happen. A crisis arose. Evolutionary drivers — limits to growth! The cells were over-populating, polluting, stagnating. The "quality of life" went down. If the Club of Rome had existed they would have warned of "the predicament of cell-kind!"

There was a "choice." To adapt to the limitations of Earth-bound energy and stagnate, or reach for new energy and grow. The "decision" was to innovate and transform.

• • •

Another amazing design innovation entered history: PHOTOSYNTHESIS — the ability to capture solar energy, "Suddenly, somewhere in an acid bubbling pool, life's lease was renewed. Even if we never find its remains, we know the nature of the originality of the new cell. Its crucial invention, probably achieved over relatively few generations, produced the first molecules of chlorophyll, the green pigment in plants which now blankets the globe."

Chlorophyll transformed the early Earth. Those first green microbes in a stifling, starving world found how to capture the energy of sunlight

to make food. With this discovery of photosynthesis, processes were initiated that two billion years later led to the fulfillment of potentialities long on life's agenda.

We can identify with the chlorophyll molecule. That fabulous process created our flesh and blood and is creating it now.

• • •

But there was a new crisis — oxygen. Chlorophyll, the technological optimist, had invented photosynthesis, a technological "fix," which produced a pollutant, oxygen, deadly to the single cells. Photosynthesis was a dangerous technology.

If there had been an Office of Technology Assessment . . . what would they have said?

How would a Congress of the Single Cells have voted on photosynthesis . . . ? It certainly did hurt the immediate interests of their constituents.

Environmentalists would have been passionately opposed to the desecration of the early Earth, the damage to the existing life forms, and the arrogance of the aspiration of life to transcend limits.

Theologians would have condemned the greedy cells for over-consumption.

Fundamentalists would have said, "If God intended us to have a biosphere, He would have built one!"

And early Earth lovers would have formed a society to "return to nature!"

None could have predicted the result of the crisis and the transformation — plants, animals, the biosphere and . . . us.

CRISES PRECEDE TRANSFORMATION

The emergence of free oxygen presented a deadly threat to the existing forms of life. This threat turned out to be a major evolutionary driver leading to the emergence of multicellular organisms able to handle the biotechnological requirements of oxygen metabolism. The limits to growth were overcome. Oxygen was used to build the new world.

The spread of life began — colonizing, migrating, transforming the "hostile" land into the verdant biosphere.

And again the pattern of crisis, limits, potential catastrophe, design innovations, a more complex technology, new forms of life and — a rise in

THE
EVOLUTIONARY
SPIRAL
CONSCIOUS
EVOLUTION
DESIGNING
INTELLIGENCE
CULTURE
UNIVERSAL
HUMANITY
1945
PHOTO-
SYNTHESIS
HUMANITY
2-3 Million Years
GENETIC
CODE
MULTICELLULAR LIFE
800 Million Years
LIFE
3-5 Billion Years
SYNTHESIS OF
ELEMENTS
EARTH
4-5 Billion Years
UNIVERSE
15 Billion Years
EVOLUTIONARY DRIVER
"Problems" which occur before every loop
DESIGN INNOVATION
The Response to the "Problems"
QUANTUM TRANSFORMATION
The leap from the past condition to the new condition
DESIGNING INTELLIGENCE
Core of Awareness running through the spiral

consciousness and freedom. Multicellular life has more awareness than single cells.

• • •

With multicellular life, sexual reproduction and scheduled death of the individual entered the process. What we, from our short perspective, think of as eternal verities are actually historical events.

Remember, the single cells divided to reproduce. Multicellular organisms were in general too complex to divide. Dr. Gorney thinks it was two paramecia:

"The later-evolved paramecia, cigar shaped with tiny oars called cilia, developed a unique new method of rejuvenation employing cooperation. The paramecia approach . . . adhere to one another. Their cell membranes open. Their nuclei draw close. Then, a remarkable collaboration occurs. The nuclei exchange half their genetic material . . ."

(Sex is information exchange, whatever the lovers think they are doing.)

" . . . they separate, revitalized, to continue dividing asexually through many more generations. In these minute creatures we find not only the essentials of sex, but also, out of the simple chemical cooperation of primitive colonial bacterial cells, the elaboration of the first animal beginnings of love. Here between simple paramecia, we glimpse the early melding of the rudiments of sex and love which is so problematic an achievement between two human beings."

The result of sexual reproduction was the diversity of animal life — flexibility, newness, variability. But multi-cellular life stealthily introduced a somber revolution. For the first time, death of the organism became a normal event in the life cycle. The clock of death was set hundreds of years ago, and is still ticking in us now, programming us to die. We have back-up systems to kill us. If one thing does not, another will. But note, sex and death are historical events. They are not necessarily eternal conditions. They entered the process together and they may pass out of the process together.

EXPECT THE UNEXPECTED/ANTICIPATE THE NEW

The dawn of self-consciousness. Imagine what it was like to awaken in the animal world with the first flickering of human awareness. We sense we are different. We are separate. We are afraid. We know of our own death and seek to overcome it. We look upward toward the stars and yearn for something more — a life beyond this life. We sense intimations of immortality — clouds of glory. We listen for voices — we receive signals from a legendary line of gods, higher beings directing us out of our existing condition. The hunger of Eve, the thirst of Prometheus — the desire to know more, to do more, to become more, arises. Future-oriented consciousness enters the scene.

We are born into a "natural" system of kill and be killed. Earth is an environment of violence. Millions of species eating each other alive. But we sense something is wrong. Killing is evil. We feel guilt at our animal behavior — and act even more destructively.

The flame of expectancy ignites. We have eaten of the first tree — the knowledge of good and evil. We feel separate from the animal world. But there is a second tree in the Garden . . . the Tree of Life . . . the knowledge of how the invisible processes of Creation work. A hunger awakens for that tree — that has never been satisfied. It is the hunger of Eve, Eve-consciousness, the fundamental motivation of the separate, creative human for immortality, for union with God. It has been driving us to reach for a higher state of being ever since the beginning of self-awareness.

In the Biblical story of creation the "Lord God became alarmed. Adam and Eve were heading toward the Second Tree! "Behold, the man is become as one of us, to know good and evil: and now, lest he put forth his hand and take also of the Tree of Life, and eat, and live forever: Therefore, the Lord God sent him forth from the Garden of Eden to till the ground from whence he was taken."

We were in a state of separate or self consciousness. Eve consciousness could not be fulfilled in immature humans.

We were expelled from the Garden. We separated from the animal world; the creature became alienated from the Creation.

THE NEXT QUANTUM TRANSFORMATION WAS US . . . HUMANITY

Two million years ago, Australopithecus prehuman. Five hundred thousand years ago, Homo erectus. One hundred thousand years ago, Neanderthal. Fifty thousand years ago, Homo sapiens. Multicellular life developed the large brain, a brain so endowed, so complex, that even now its capacity has barely been tapped. We are still a young species.

We don't know the evolutionary drivers that triggered the emergence of Homo sapiens from the animal world. According to Richard Leakey, in his latest book *Origins*, the survival factor that led to the dominance of Homo sapiens over the other big brained creatures — Neanderthal and Cro-Magnon — was the capacity to cooperate, a food sharing economy, heightened sociability. The core of the behavioral difference between

Australopithecus and ancestral humans was that Homo sapiens established home bases and shared food.

The critical human innovation, comparable to the genetic code and photosynthesis was CULTURE — language and symbols — the capacity to communicate information exo-genetically, outside the genetic code.

The now familiar pattern was at work again. The human brain, the most complex synthesis ever developed on Planet Earth brought with it something utterly new — Humanity. And with Humanity a design innovation — culture — another giant leap in consciousness and freedom.

Self-consciousness must have been unstable in the animal world, just as cellular consciousness was unstable in the pre-life seas of passive molecules. Eventually self-consciousness stabilized and became the new norm.

• • •

The monumental struggle began to overcome the mammalian condition — we have always wanted to go beyond the animal limits of eating, sleeping, reproducing and dying. All of human culture can be interpreted as the effort to transcend — that driving dream of the human psyche to go beyond the bonds of animal life. Through tools, language, religion, art, agriculture, and now recently through science, industry and technology, we strove for something more, something new. We transcended our limitations through learning how the material world works. We never settled for comfort. There were always among us the pioneers, the transcenders who reached for the next step.

We humans, like the cells, became successful. We reorganized other life, utilizing resources, converting them into our own systems — migrating, increasing in intelligence as we faced new challenges and gained mobility, spreading around the globe.

Recently societies became over complex, over-extended, unmanageable. We sensed the finiteness of our planet, the impending limits to growth, the stagnation and pollution of an over-populating species. The situation began to go critical again. Incremental change was leading to catastrophe or the creation of something new.

• • •

1945 marks the beginning of the next phase. The design innovation, CONSCIOUS EVOLUTION, is preceded by crisis! The atomic bomb — an evolutionary driver for the human race.

Our still selfish species, in a state of separate or selfish consciousness was stunned by the Promethean gift of nuclear fire . . . a power of creation or destruction on a planetary scale. The collective egos of nations and races were signalled to cooperate or die. However, if we chose to innovate and transform, as all life before had done when confronted with crises, if we harmoniously applied our new potentials in astronautics, genetics, cybernetics, physics and the understanding of our own nervous system and brain . . . we would transcend historic limitations quickly, organize and integrate at a higher, more complex level . . . and become something more, something new.

The next transformation had started. There was no turning back. The flame of expectation that had become conscious at the dawn of Humanity was guiding us toward the Tree of Life.

Human history has been the story of this desire to become more than animals, to be like the gods and live forever.

The evolutionary significance of the brief period from early Homo sapiens, fifty to one hundred thousand years ago, to early Homo-universalis, at the end of the 20th century, is that during this brief period of the "Fall", we sensed ourselves separate from nature long enough to intuit the presence of a Creator, a God-force with whom we wished to unite. We also learned the first lessons of how the Creation works — how the material world operates — the technologies of Creation. The period of self-consciousness permitted us to become objective, to stand outside the Creation long enough to develop a conscious relationship with the Creative Intelligence through spiritual experience and to gain empirical knowledge of the Creation through science and the rational mind.

Now our generation is approaching the Tree of Life — the knowledge of how the invisible processes of Creation do in fact work — the brain, the atom, the genes, the cells, the stars. We are ascending from self-consciousness to cosmic consciousness, from creature to co-creator, from Earth-only to universal. We are about to become like the gods, partners with the evolutionary process — or perish.

There is no way to avoid the power of our knowledge. We cannot unlearn what we know, or stop knowing more.

This is a cosmic drama! The gods themselves may be holding their breath to see if we can fulfill our hearts' desire.

• • •

However, lacking an evolutionary perspective, many on Earth are traumatized. They announce that the end of civilization is at hand. Progress leads to power leads to destruction. People are the destroyers. "We've met the enemy and he is us!" Our self-confidence disappears. We lose our image of the future — a positive vision to attract our energies.

Materialistic scientists proclaim the inevitable heat death of the Universe — the running down of material energy to a state of maximum disorder. God is dead. Aspirations for a greater freedom, consciousness, purpose, immortality seem ridiculous. Freudians interpret all higher aspiratins as repressed sexuality. Darwinians reduce us to nothing but intelligent animals, fighting for survival. Existentialists write about us as absurd accidents in a random, purposeless universe. The modern arts herald the end. No future. No hope. They see what is dying, not what is being born.

• • •

From an evolutionary perspective we see the familiar pattern. We need not be afraid although we do need to be alert. Evolutionary drivers, or threats, are essential to overcome inertia. Crises always precede transformation.

The energy crisis, the population crisis, even the nuclear crisis, are essential evolutionary drivers stimulating innovation and transformation. In that sense they are "good." They are causing us to change.

And in fact there are vital signs of response to the crises. The attractive force that runs through the spiral organizing atoms, cells, and the brain is operating on our planet now, working to integrate us into a new whole. It's absurd to think the force of Creation would suddenly turn off with us! The pattern of complexity/consciousness is still active, breaking into our awareness and action now.

We are not doing this alone. The force of Creation is with us.

The United Nations is attempting to bring order to the selfish nations. Energy solutions are emerging — renewable energies of all kinds are being developed, with devoted lobbies and activists dedicating their lives to the achievement of non-polluting abundant energy for the people of Earth.

The environmental movement has become world-wide in the last ten years — since we all saw Earth as one body from space. Every town, village and city has its passionate environmentalists protecting, conserving, demanding ecological care of the web of life.

Movements of participation are irresistible. No state or institution can resist the tide of Humanity demanding to become an active part of the system — blacks, women, ethnics, the old, the young, the poor, even the rich are demanding a piece of the action.

Movements of expanded consciousness are breaking the bonds of the institutional churches, seeking more direct experience with the spiritual dimensions of life.

Movements for human potential, affirming the innate goodness and creativity of all people, are superceding the Freudian and Behaviorist views that we are nothing but the sum of our lower drives or our glandular reactions. We are perceiving ourselves as beings with unlimited capacity to grow, to transcend, to overcome self-imposed limits of our physical, mental and spiritual past.

Movements for civil rights, for human rights, though slow and painful are on the rise. No matter how tyrannical the state or how brutal the prisons, voices of freedom are communicated worldwide. They can never be silenced again.

And at the growing edge of human ability, in the new sciences and psychologies — astronautics and genetics, the biological revolution, the cybernetic revolution, the brain-mind revolution — in the exploration of the extrasensory and the extraterrestrial dimensions, we see the outlines of the genuinely new.

Yes, the vital signs are there and we are beginning to respond.

• • •

Some limited optimists believe we can survive through adaptation to the existing limitations of an Earth-only species. But this approach ignores the new potentials. It's like seeing the future from the perspective of the womb.

Imagine an erudite fetal scientist at the seventh month. "Now esteemed colleagues, our indications in this seventh month of this cycle are that by the eighth month there will be severe overcrowding and famine. By the ninth month we will have run out of fossil fuels. Mass starvation will begin. In the tenth month pollution, overpopulation and social revolution will intensify to catastrophic proportions, by the twelfth month, we will have massive die-offs . . . therefore, we must stop growth now, redistribute existing resources, conserve, preserve, manage, plan and accept our limitations."

Like most experts our fetal scientist is unable to see the new, since it has never happened before. It is not obvious to a cell in an embryo that in the ninth month birth occurs. It is not obvious to a member of an Earth-bound planetary body that a certain stage of growth, universal life breaks through.

Our media, our planetary nervous system, is infantile. Like a baby's nervous system it is reporting pain, irritation, danger . . . then it dulls us to sleep. It is sensitive to breakdowns, but not to creative breakthroughs. People in general are not aware of their planetary potential growth. And for lack of this awareness, they sink into despair, cynicism and withdrawal.

• • •

Only from the perspective of fifteen billion years of history can we understand the process of change, take courage and be joyful. It is the only appropriate context to understand the potential of our future.

Could molecules have predicted cells? Could cells have predicted us? Perhaps not.

Can we predict the next step? Perhaps we can. The process is prophetic. The process is predictive. Remember, we have learned to expect the unexpected, to anticipate the new.

Chapter 4

THE PRESENT
The Unwritten Script

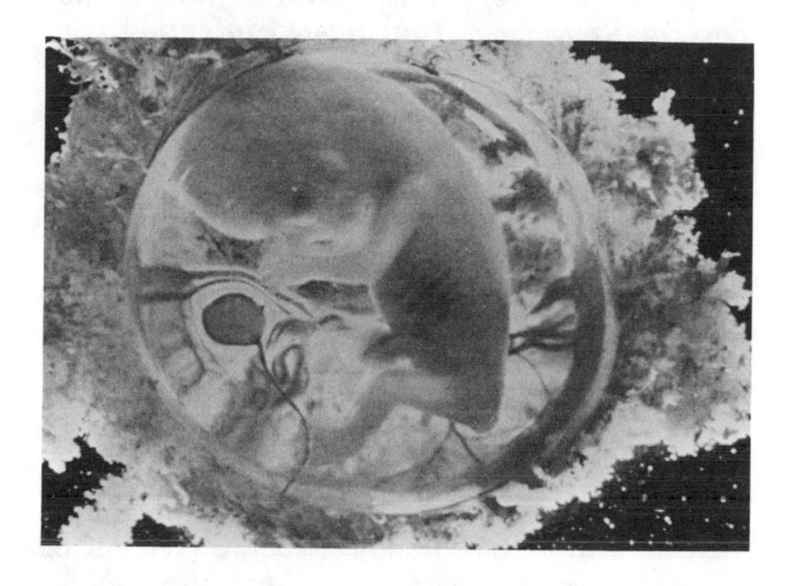

The physical penetration of the infinite has begun with our space program. At the same time that we are reaching deeper within ourselves for contact with reality, and recognizing the finiteness of our mother planet, we are becoming young extraterrestrials. Newness! We have learned to leave the planet alive.

Using your imagination, let's burst the blue cocoon of Earth, crack the cosmic egg and lift ourselves with a space rocket. We are in space looking at our little Earth, just as the astronauts did. We are all members of that one body, which we see floating before us.

Weightless there in outer space, we ask the Universe a question: What is happening to us? What is our story?

The answer echoes through the vastness of the womb of Creation: "Our story is a birth. It is our own. It is the birth of Humanity." It's happening now. What the saints and seers predicted is true.

We are one body born into the Universe discovering awareness of our creative intention.

• • •

Each of us is a vital part of our planetary body. We can feel ourselves struggling to coordinate throughout the world. We are running out of energy. We are gasping for breath. We are feeling the pain of oneness coursing through our body as our communication system flashes the pain of one to all. We are reaching into the Universe for new sources of energy, new resources, new space to grow. And as we reach out our cells are linking, our separateness is fusing, our body is becoming whole. The reach into space and the link-up on the body of Earth are one gesture in cosmic time.

Our whole body floods with the passion of oneness, the excitement of newness, the fascination with our future on the threshold of revelation of who and where we are.

The Cosmic Child smiles. We are that child. The whole planet sings a song of celebration — for its own birthday. The whole planet hears the same sounds, feels the same joy, thinks the same thought. We are Humanity, one body, born into the Universe.

• • •

We lift our eyes together beyond our world and sense the presence of other life — benign, loving, longing to communicate with us. We do not know the language yet. We are still too young. Our nervous system has not quite linked up — but we will learn to speak the universal tongue. Like a newborn baby we are surrounded by life but still too young to know it.

We remember our cosmic past, our Self's conception, gestation and birth. We relive the evolutionary spiral. It is our pre-natal past. We can hear the music of Creation — intensifying at every quantum transformation, building slowly, speeding, becoming cacophonous at each point of transformation, then resolving the crises in great harmonic chords that blend all the themes into a triumphant choir of rising consciousness and freedom.

We dance the Creation to the music of the spheres. The conception, the explosion, the formation of our Earth, the awakening of the first cell, the organization of the multicellular creatures — plants, animals, birds, then Humanity breaking into the spiral with awareness of more to come — and now us, born into the Universe, ready to become that which can be.

The evolutionary spiral is the story of our birth. We have good news to tell.

From the deep dark womb of evolution,
Humanity has been born.
Our eyes and ears are opening,
We listen and understand.
As one body we study the new description
Of the forces of Creation
Gleaned for us by science.
In the light of our dawning awareness, we seek
Deeper contact with the Creative Intention
And the role we can play in universal affairs.

We are the generation born when Humanity was born. We are the first generation to be aware of ourselves as one, responsible for the future as a whole. Our capacities as humankind are infinitely greater than as individuals, separate and alone.

• • •

We fall in love with ourselves as one body. It is impossible to love our North and not our South, our East and not our West. We feel parental to ourselves and forgive our whole species. We are a cosmic infant. We have faith in the potential of every newborn child, no matter how self-centered it is at birth.

We have been taught to forgive one another, one by one. Now we are taught to forgive ourselves as a whole.

We have been taught not to judge that we be not judged. Now we are also taught not to judge the world.

We have been taught that God loves each of us. Now we are teaching ourselves to love the whole world.

We love ourselves as a parent loves a newborn child, bestowed by the process of Creation with the potential for universal life.

We bless and affirm our new powers. They are natural capacities of a universal species.

We are making the birth transition from Earth-only to universal life. We must heal our wounds and grow. The last loop on the spiral is the loop of our birth. When we grow through the transition we will be Universal Humanity.

• • •

What is the design innovation in our quantum transformation comparable to the genetic code, photosynthesis and culture?

CONSCIOUS EVOLUTION is the word we place on the evolutionary map. The innovation needed now is *to understand the magnificent processes that created us and purposefully cooperate with them in planning and designing our own future.* It is essential for survival. We know we have the means to destroy our whole planet. We also know we can create a future of unlimited possibilities. No generation has ever been asked to learn how a planet operates. We can have compassion for ourselves. There are no "operating manuals for spaceship Earth," as Buckminster Fuller says. There are as yet no schools of Conscious Evolution.

We are learning from the text of fifteen billion years. How does mother nature do it? How does she maintain a balanced ecology? How does photosynthesis work? How does life replicate? How does the atom work? How do we build new worlds? How do we create more complex whole systems that have greater freedom and consciousness?

• • •

What are some of the tools of Conscious Evolution? A key is *synergy* — learning how to form whole systems, and how separate parts work together to form a new whole which is greater than and different from the sum of its parts. Our body is a synergistic whole — far different from the sum of trillions of unique cells.

At every quantum transformation in the spiral a new synergistic whole system was achieved: the cell from molecules; multicellular organisms from single-cells; the human brain from the animal. Synergy is a natural tendency, one of nature's traditional methods of organization.

Now, at our quantum transformation, on every level of life, we are trying to learn the art of synergy. Psychologists teach us to become whole persons, integrating body, mind and spirit. Wholistic health care practitioners integrate specialized medicine. Environmentalists study how the whole biosphere functions as an interrelated web of life so we can consciously maintain it. Community builders, organizational managers, citizen participation groups, cooperatives of all kinds seek consciously to create environments in which each person joins creatively with others within a whole. Universities and bureaucracies struggle to overcome the sterility of separate disciplines and departments which are unable to function as a whole organism. New Age spiritual groups help the human will to cooperate with the divine will, freely. And space scientists, architects, environmentalists and sociologists design plans for whole new worlds.

THE SYNCON

There are many meetings, gatherings, encounters which bring people together to interact as a whole group. In the Committee for the Future we practiced synergy in twenty-five conferences call Syncon,* for synergistic convergence. We literally built a wheel and invited people who "hated" each other to join together to see how each could realize his or her own needs for growth in the light of the whole social body. The first meeting took place at Southern Illinois University in 1972, at the height of the Spirit of Revulsion of one group of another. Environmentalists hated technologists. Blacks hated whites. Women hated men. The young hated the old. Radicals hated conservatives. Scientists hated

*For further information contact: Futures Network, 2325 Porter St., N.W., Washington, D.C. 20008

psychics. The planetary baby had an allergy against itself! One part of the body got hives at the very thought of another!

We asked these groups to come together to see if each could meet their own felt needs, without damaging another sector, when placed in a context that included the capacity of the whole system.

People met in the Wheel in task forces according to basic functions: Environment, Technology, Production, Social Needs, International Relations, Government. They were asked to state their own goals, needs and resources, just as they saw it; selfish as they wished to be. At the "growing edge" we invited geniuses who were developing new potentials: in biology; in the information sciences; in the psychologies of growth; in the physical sciences, in political/economic theory. We asked them to piece together the existing potential of Humanity as a planetary species. Artists, represented symbolically as the skin of the whole, were asked to help us envision what we look like, where we are and what we desire to become. Finally, a task group called "Unexplained Phenomena" included people who were having unexplained experiences — telepathy, clairvoyance UFO contact, spiritual guidances of all kinds. The Syncon Wheel is a people mandala, a microcosm of the social body, providing a "ground of the whole" in which the parts of the social body can find their deeper relationships to each other.

We placed television cameras and monitors in all sectors — an internal nervous system for a social body, so people could sense themselves as part of a whole body, even though each was focusing on a specific part. Hundreds of hours of videotape now exist documenting social transformation through synergistic action.

Every evening we played back a television "New Worlds Evening News" highlighting convergences, sudden agreements, creative linkages, instead of the usual emphasis on disagreement, breakdown and eccentricity. There were special cultural events each night designed by the artists.

In the center of the wheel we placed a "syncon-sole" — a little social mission control with many small TV screens showing the work of the various groups at the same time, demonstrating Robert Frost's insight that people "work together whether they work together or apart."

Gradually, through a process of scheduled mergers, corollary but apparently conflicting functional groups met — environmentalists with technologists, business people with welfare mothers — looking for common goals and matching needs and resources. At the "growing edge" we asked the groups working on new potentials to report to the whole what it would be like if the emerging capacities became operational at once . . . if everything worked!

A feeling of awe took over. Linkages among conflicting groups spread like ripples in a pond while the reports from the growing edge stimulated a self-revelation of a social body to itself. We caught a glimpse of what we already can be — so much greater than what we are actually doing. Awareness of our common potentials infused the participants. Something amazing began to happen. Like the macro-molecules becoming a cell, we were fusing into a new whole. We were synergizing!

• • •

At the end of the *Syncon*, the walls that divided the various task groups in the Wheel were physically removed — leaving an "Assembly of

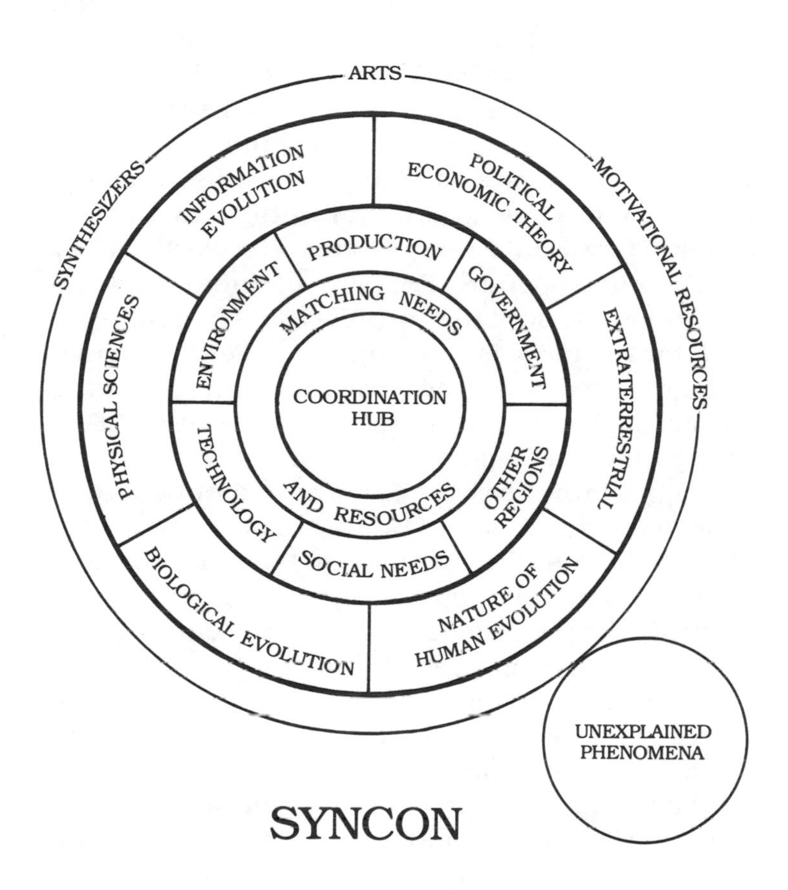
ARTS
SYNTHESIZERS
MOTIVATIONAL RESOURCES
INFORMATION EVOLUTION
POLITICAL ECONOMIC THEORY
PHYSICAL SCIENCES
EXTRATERRESTRIAL
BIOLOGICAL EVOLUTION
NATURE OF HUMAN EVOLUTION
PRODUCTION
ENVIRONMENT
GOVERNMENT
TECHNOLOGY
OTHER REGIONS
SOCIAL NEEDS
MATCHING NEEDS
AND RESOURCES
COORDINATION HUB
UNEXPLAINED PHENOMENA

SYNCON

the Whole" in a circle with no divisions. It was supposed to be the last session, where unresolved conflicts could be worked on by the group as a whole, aiming for solutions wherein the needs of one group did not damage the needs of another. We were to strive toward "all win" solutions.

But as the walls came down, people were drawn to each other by an irresistible force. They began to celebrate, dancing, weeping, hugging each other. They would not return to the old order. They would not go back "to work." They would not separate. For the moment they no longer cared about the "problems." The problems had, for the moment, disappeared. Fear, distrust, antagonism dissolved in the overpowering warmth. People wanted to love each other! Empathy increased and drew the shy, the skeptical, the angry into the circle. Longhaired radicals embraced conservatives in business suits. Students clasped hands with professors. Psychics and scientists would not stop talking. A bonding was occurring. People said they wanted to "do it again," to "come together again." The language became "suprasexual." The parts were being attracted into a whole system which enhanced each member and decreased none.

The separation was over for that instant. The energy of repulsion and hate turned to attaction and love. The "problems" which divided the participants three days before had changed.

Synergy was at work. The same force that organized every quantum transformation was bringing humans together in a social whole.

• • •

A remarkable social innovation was occurring. Suprasex! Remember that sex was "invented" at the beginning of multicellular life millions of years ago. It is information exchange at the genetic level to produce the diversity of the species.

Suprasex may be a step toward the next form of social union.

People fuse, not their bodies to procreate babies, but their ideas, aspirations, and intentions to co-create the future. They are not merely "helping" each other, anymore than a man and woman are "helping" each other have a baby. It's mutual creation, not charity!

At *Syncon* we were enabled and empowered by that form of social union to create something that none could do alone. Our potentialities as unique individuals were being drawn forth by the needs of others for what each does best. It felt so "good."

When people who had not experienced *Syncon* ask, "Isn't it idealistic to expect individuals to cooperate as a whole group?" the response was, "Is sex idealistic?" No! And neither is suprasex. It's inherent in the nature of reality for people to be attracted together to become more. What is needed are many non-threatening wholistic contexts — assemblies of the whole — throughout society, with a modicum of process for interaction among conflicting groups. Nature will do the rest. We are inexperienced at suprasexual union, but we learn quickly! It's like putting two adolescents in a dim room. They might not know what to do at first. . . but they learn!

• • •

Nature's secret intent is union for the creation of new life. In our efforts to "come together" in loving whole groups, we are not working alone. The laws of attraction are working with us. The power that organized the Universe is organizing us! When we feel the excitement of fusion pulsing through our bodies, drawing us to other people, it is the Creative Intention, the godforce, the irresistible magnet.

This excitement is essential. A bored species will never make it! Nature put the capacity to be aroused at the center of the act of physical creation. An unexcited man cannot procreate. And an unexcited group cannot create a desirable future. The sacrifice, the hard work, the patience that the nurturing of life calls for must be buoyed up by joyful union. We can trust the "compass of joy." When we are participating in the formation of a new whole, joy enters our life. Connecting is pleasurable. It is the essence of growth.

When we fuse as a group, our consciousness begins to rise to a new kind of spirituality. We crave to stay connected. This deepening togetherness stimulates a form of consciousness that does not yet have a name. Evolutionary consciousness, co-creative consciousness, may be appropriate. It is a synthesis of mystical and secular awareness emerging in this super-rich information environment under the pressure of convergence and growing responsibility.

Erich Jantsch calls it "syntony," or superconscious learning. We sense that there is a transcending evolutionary process which runs through the core of the spiral, a Designing Intelligence operative throughout the Universe, which we humans did not invent. "It" invented us!

• • •

As group-connectedness and synergy increase, like a more complex radio receiving set, we are capable of attuning to the signals from the process more clearly. Perhaps through some resonance or vibratory phenomenon, like the DNA communicating the plan to the RNA messengers which gather the proteins to build the cell, we are receiving information from a planetary information code instructing us how to build a planetary whole, a new synergistic civilization greater than the sum of the separate parts.

What has been experienced in the past as mystical insight — such as St. Paul received on the road to Damascus — sudden, traumatic, blinding enlightenment, almost too powerful for the nervous system to endure, is becoming a new shared norm — the democratization of awareness. As we interconnect as a planetary whole, each person is becoming a clearer channel to the "Source," the Process of evolution, God.

The complexity of planning for a planetary body is too intricate to be understood by rational, empirical knowledge alone. We also need super-intuitive knowledge to survive.

It feels as though the organic process of planetary transformation is being orchestrated at a higher level.

Paradoxically, as we become more responsible for the process, we are also more response-able. We are receiving deeper super-intuitive guidance.

We "know" that our physical bodies are hungry, sick, tired, etc., by complex means. Through our extended planetary nervous system (television, telephone, satellites, computers) and through increased "syntony" we begin to "know" when members of our extended planetary body are threatened — without having to read hundreds of books to understand. We feel the direction of health for Earth in our own blood and bones. Pollution is not "out there." It is happening to us. If people are hungry anywhere in the world, we feel it. If people are fighting each other, it's a fever in our body. All pain is ours. Mindless destruction of any part of nature is destruction of us. If our space programs are cut off through lack of awareness, it's our body whose future is suppressed. It's our energy that's cut off.

• • •

Evolutionary consciousness is active. We want to act upon it. Networks of awareness are forming throughout the world. They can respond with the speed of shared intention to act on behalf of a planetary need.

Perhaps signals from the volutionary process intensify at the times of quantum transformation when the organizational pattern is being changed.

How many of us would like to rest, to take "a breather," to settle in for a while into a secure and comfortable position . . . when we are kicked from behind and attracted ahead as though by some relentless magnetic force.

How many of us are receiving signals, voices, inner guidance? How many are becoming functional pioneers, inventing new vocations to meet new needs?

When you feel the next painful thrust to grow, think that it may be the Universe itself growing through you.

• • •

Syntony, synergy, suprasex are some tools of conscious evolution. Increasing attunement through interconnectedness, plus growing scientific information of whole systems, plus intensifying suprasexual attraction and breaking down barriers of repulsion, are some of the elements of the design innovation we call Conscious Evolution.

Conscious Evolution is probably a natural capacity of a universal species at the dawn of its entry into its cosmic environment. Like a baby, a cosmic planetary being needs to learn how to do for itself what its Mother Earth did for it in the womb — to coordinate, to eat, to eliminate, to walk, to be responsible, to be curious, to learn, and eventually to become independent of mother's care.

Quantum leaps happen relatively fast — remember the chlorophyll molecule and the rapid emergence of the big brain. In just a few generations it's possible to make the critical crossover from the old to the new. Our process of transformation is already well under way. We may live to see what is coming within our lifetime.

• • •

Let's piece together what we already know we can do — or almost do. Let's imagine what it will be like when everything works — when we are using our capacities as a species harmoniously. Imagine the difference between a newborn infant screaming, frantically seeking its mother's breast, thinking of nothing but its own needs . . . and that same child several months later, smiling, crawling, eating, learning, loving.

Remember, we are taking the perspective of a more mature species looking back upon ourselves.

What is it like to grow up? We need not be afraid. It's probably happened many times before in this abundant Universe of ours.

Chapter 5

THE FUTURE
Previews of Coming Attractions

Now imagine that the sleep of the womb is over. We are growing up, becoming fully human.

We experience infinitely expanding awareness . . . deeper and deeper interrelatedness. The veil of matter disappears. Science sees bodies as energy in motion. Reality is more like thought than a thing. Empathy increases. Intentionality expands.

We make closer personal conscious contact with the evolving core of the spiral. The Designing Process that organizes the Universe breaks into our awareness as we ourselves begin to organize on a planetary scale.

Centuries ago great religious avatars came to this planet in a state of cosmic consciousness — the Indian seers, Akhenaten, Moses, Buddha, Jesus, Mohammed. They sensed themselves as part of the whole Universe attuned to the directing intelligence of a Creator guiding them onward to universal life through love of each person as a member of their own body. All taught the same ethic of love, the same faith in God, and the same promise of eternal life beyond the bonds of this planet and this mammalian body. All knew directly they were integral parts of the Creator. But few "ordinary" people could share their experience.

COSMIC CONSCIOUSNESS

Now millions are attempting direct personal contact with a deeper creative reality through meditation, prayer, yoga. We want to feel the Force. We want to work with the Force. We want to become the Force.

Christ / Buddha / Mohammed / cosmic consciousness becomes the new norm. Just as self-consciousness was unstable in the days of early Homo sapiens, isolated in the animal world, so cosmic consciousness is unstable and still rare among the billions of minds slumbering in the sleep of separation, trusting as real only what their five senses report.

Cosmic consciousness will be established. The limited period of self-consciousness will be over. Evolution has been, still is, and will be a consciousness-raising experience.

Everyone who makes it through the quantum transformation will be in a state of cosmic or wholistic awareness.

It's as natural as the evolution from animal to human awareness.

In retrospect, we may not even remember what it felt like to be in the prison of the "self-concentration camp." We have endured the pain of feeling separate, unrelated to each other and to the universal process as a time of learning and preparation for the next phase, to take our rightful place in universal affairs as young co-creators. We have been to the school of Homo sapiens for thousands of years, and are about to graduate to the next class of being—Homo universalis.

COSMIC ACTION

Cosmic action beings, the natural complement to cosmic consciousness. They are the inner and outer aspects of the same capacity. Our infant space programs are the early exercises of a species about to enter the designing process consciously. Co-creatorship is real, not a metaphor. We identify with the Creator as well as the creature as we become more creative. We are planning new worlds! Literally. Now.

The vision of physical cosmic potential was first conceived in the minds of a few lonely men — Goddard in the United States, Tsiolkovsky in the Soviet Union — as an aspiration to transcend the limits of Earth-bound life. From that drive for physical transcendence rose the magnificent Apollo, trailing clouds of flaming glory to the moon — the most perfect, complex act of creation ever achieved by humans. We penetrated the infinite and become young extraterrestrials.

Now begins the period of consolidation and usefulness. The Earth is no more closed than it is flat. We live in an indivisible Earth/space environment. The space shuttle is the first bus to the infinite. It initiates the era of space industrialization, satellite communications, energy from space and permanent occupancy, establishing a new environmental niche for the human race.

A new era of abundance begins. We become creators of new resources rather than managers of scarcity. Just as limits to growth are reached on our Mother Earth, we gain access to cosmic abundance, stimulating new productive employment, new incentives for international cooperation and non-inflationary, beneficent growth. The ancient Earth-bound plague of scarcity is over. We have invested in the infinite and the "return" is astronomical. Space is a medium for growth. Astroculture begins.

Within twenty years we achieve the next evolutionary milestone: the establishment of the first non-terrestrial resource base. We learn to build "new worlds" in space out of lunar materials and solar energy — sunlight and moon dust! We no longer have to return to Earth, down the 4,000 mile gravitational well, to haul up material. We are free!

We now have the technological ability to set up human communities in space: communities in which manufacturing, farming, and all other human activities can be carried out. We are Earth interdependent and eventually Earth independent.

• • •

This step is comparable to the fish establishing themselves on the dry land. The tremendous technological advance from the sea-dependent amphibians to the sea-independent reptiles paved the way for the multitudes of land and air species, colonizing the "hostile" dry land leading eventually to us. Life's confrontation with the challenges of the "new" environment, dry land, driven by the urge to transcend, was the forcing function for transformation, diversity, newness, and a leap in awareness and freedom.

Expanded consciousness was developed

along with expanded bodily capacity to act, and new technological skills. Billions of new bodies, hundreds of millions of new species explored every nook and cranny and developed the hands, feet, eyes, brains and nervous systems that we have inherited, and which now are taking us toward cosmic life.

The result of our capacity to live permanently in space will be unprecedented evolutionary di-

versity and eventually the creation of a genuinely universal species. The first child conceived and born in outer space will be a step toward a new being.

The amphibians and the astronauts have much in common. Both are transitional links, creatures animated by an urge beyond their comprehension, giving birth to an unpredictable next stage of history — yet potential in us now — prepatterned but not predetermined.

Even now plans have been drawn up by the National Astronautics and Space Administration

(NASA) and by private groups, to build designed habitats in free space, at a stable orbit between Earth and moon which can produce energy and products for Earth and will be capable of replicating themselves, using nonterrestrial resources.

By the year 2000 we could be an Earth-space species, building many worlds. And beyond that, we are free to explore the unknown.

Krafft Ehricke calls this step the "androsphere," the human sphere. We inherited the biosphere. It was the creative innovation of the chlorophyll molecule that developed photosynthesis and of the billions of organisms that produced our "natural" environment. Now through Conscious Evolution, it's our turn to create a new habitable environment — in the great fifteen-billion-year tradition. The human brain is beginning to develop the human sphere, beyond our biosphere, the next "natural" step in evolution.

This capacity will lead to the spread of life throughout the Universe. From the perspective of the solar system, seeds of life are leaving this planet and falling upon neighboring barren planets. Earth is bringing life to the solar system. We are "plentipotentiaries of life" in the Universe, as Roger Wescott says. Mother Earth is reproducing herself through her offsprings, the cosmically-attracted peoples, as she is healing herself through her Earth-loving peoples.

• • •

The limits to growth we momentarily feel are not ultimate, but a terrestrial reality which serve as evolutionary drivers to stimulate us both to conserve on this planet and to reach for renewable and inexhaustible resources and living space.

Our problems are the problems of success not failure. It is not sickness but health that is attracting us into the Universe. Growth is natural. Birth is natural. And there is no energy shortage in the Universe.

Building new worlds on Earth and in space are not separate tasks. They are one. We will stop exponential growth on this planet naturally as productivity increases. Population growth will fall as material well-being increases. The need for large families will disappear throughout the develop-

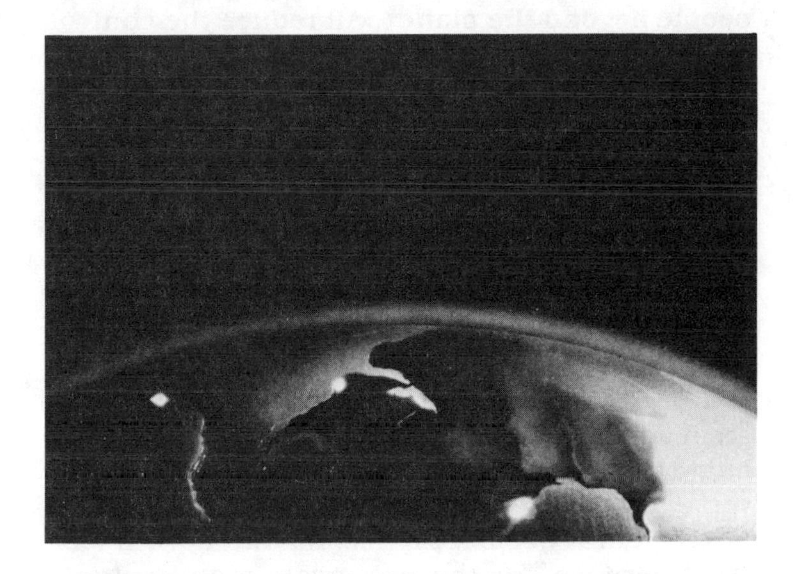

ing world as has already occurred in developed countries.

Small is beautiful on Earth. The next industrial revolution will be in outer space, beyond the biosphere. Large structures are appropriate for space.

Under the pressure of increasing planetary interdependence and complexities, centralized bureaucracies break down. Smaller, self-organizing, decentralized communities blossom. The dinosauric centralized systems evolve into communication and coordination centers responsive to the vital needs of all parts of the planetary system.

Perhaps the last great function of existing centralized power, as in the United States and the Soviet Union, is the establishment of the first productive foothold in the Universe. After that step, the higher frontier will open to cooperative, free enterprise and self-selected groups of pioneers. The opportunity for new life styles, new wealth, new knowledge gained by highly motivated people beyond the planet will reduce the control of Earth-bound governments everywhere.

The developing world can participate in the resources of the new frontier without the need to go through the "old fashioned" industrial revolution based on fossil fuels and diminishing non-renewable resources. Through the full spectrum of appropriate technologies — from labor intensive simple tools through complex space technologies, the whole world can meet the needs of all its people. One hundred percent of Humanity can be a success, says Buckminster Fuller, the great advocate of our abilities, not only to prevail but to transcend . . . to become citizens of the Universe.

The Earth becomes a Second Garden, our cultural heritage, for the preservation and cultivation of works of art and nature. We conceive of this Earth as both a beautiful place to live and a point of departure.

As Earl Hubbard writes in *The Search Is On*:

> "If we are a migrant people, if we did choose to move out into space, if we are seeking new frontiers, then our communication centers will respond to this activity. New enterprises that we cannot dream of now will develop. Life on this Earth will be as a center of operations. Our business will be

> the business of evolution — problem solving, developing new techniques, and making of this Earth a genuinely beautiful place to live. It will be the place where culture is not only conceived but sent forth."

• • •

We will preserve the endangered species. Over 90% of all species were extinct before Humanity.

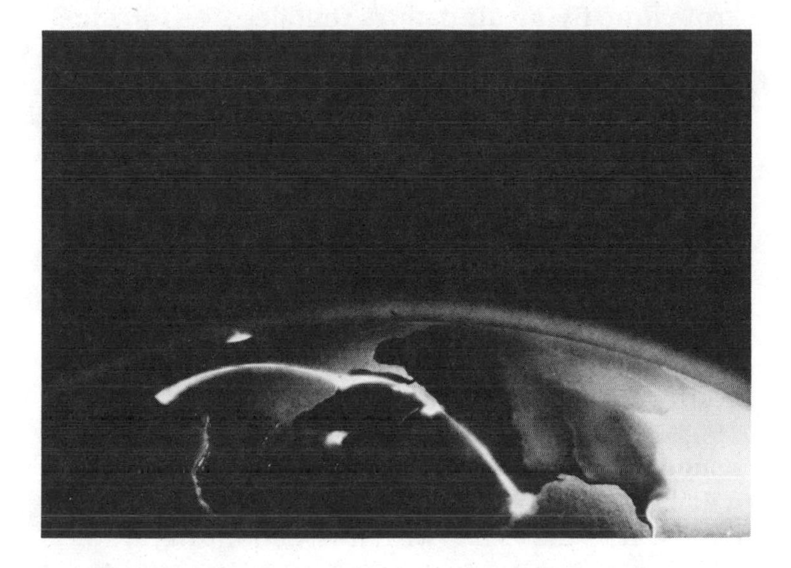

We are the first species to have the ability or the inclination to try to save others from extinction as evolution moves on.

Evolution is life-oriented. It is "good" but it is not kind. It does not care for the feelings of individuals. Humanity is the first species to introduce humaneness into the process of transformation. The cruelty lessens, suffering and sorrow diminish as the sensitivity, capacity and empathy of Homo universalis grows. We care for all individual beings and all of nature as part of one body. Christ-Buddha-Mohammed-Cosmic consciousness is the "new norm" . . . just as those great precursors of Universal Humanity predicted. (Remember, cosmic consciousness and cosmic action are happening concurrently, reinforcing each other. We are evolving physically and psychologically, just as a newborn child.)

As Earth is beautified anew, so the new worlds in space will be beautiful. Perhaps not at first. All new forms are awkward at the beginning — the first land creatures that became the gazelles, the grotesque flying dinosaurs that became the sleek birds, the hairy pre-humans that became Venus de Milo and the Mona Lisa, the first airplane that became the silvery jet. Everything that evolves and endures becomes ever more beautiful. Evolution is a beautifying experience. Every bird, every leaf, every stone, everybody that continues to exist is beautiful. Aesthetic harmony is intrinsic to life. Therefore, those new worlds in space which endure, will also be beautiful — perhaps a beauty so new and inconceivable to our Earthbound eyes, that we cannot imagine it at this early stage. We can trust that what is able to survive in space will be as beautiful as the rest of nature. It will be nature. Nature is not only what already exists. It is also that which is becoming, what is being created by creatures who inherited the brains to become co-creators.

The technology of creation stretches from the genetic code to the computer, from the eye to the telescope, from the biosphere to the androsphere. The continuum of technological innovation continues with our innovations. We are the newcomers to the process of building bodies to transcend limits. Our skills will improve.

COSMIC COMMUNITY

Next, imagine that cosmic community which will be developed as a natural component of cosmic consciousness and action. To survive on an interdependent planet and in new worlds in outer space, we will learn to cooperate at a deeper level. In the space communities, we will have the opportunity to design new models of society where social interconnectedness, interdependence and synergy are consciously built into the system from the beginning.

Cooperation is essential for survival. Pioneers

will work together — or die. We saw the necessity for this form of cooperation with the Apollo astronauts. One fight, one lie, and the mission might have been destroyed for all. Every virtue taught on Earth, but so often ignored, will become essential — maturing human behavior to a new ethical norm. Honesty, excellence, patience, trust, love of each person as ourself, becomes normal in communities which survive in outer space, and eventually on Earth.

The factor that selected Homa sapiens over other early humans such as Neanderthal and Cro-

Magnon was, apparently, the superior capacity to share food and cooperate socially. The same ability at the next level will be what is selected among early Homo universalis. The better a group can collaborate, the more likely they will survive to reproduce themselves socially, that is, by replicating new worlds in space. The successful universal pioneers will build the daughter communities which split off from the first model colonies.

The constant newness provided by the new frontier, the excitement of the unknown as a daily companion, will stimulate the intellect, perhaps carrying into adulthood the curiosity and boundless learning capacity of the child to whom all the world is new. For an evolving, self-designing community in space, every day is genuinely different. Deadly boredom, which instigates so much pettiness and loss of innovation in bureaucratic institutions, may be overcome as we enter a continuously evolving, or co-creative, way of life in the infinity of the Universe.

• • •

Why is it that no society on Earth has achieved its own aspiration of community? Beyond a certain scale, injustice and inequality always creep in to destroy the dream. Yet the dream is the same for most cultures:

"Love thy neighbor as thyself."

"Do unto others as you would have them do unto you."

"From each according to his ability, to each according to his need."

"All men are created equal."

Can it be that in the early human stage of self-consciousness and material scarcity, the dreams of the ages for loving community could not be fulfilled? They were premonitions of something to come. Can the missing factors have been cosmic consciousness — the sense of interrelatedness and attunement to the Designing Intelligence — coupled with cosmic action — the ability to produce abundance? These are the natural abilities of a universal species. Perhaps we could not achieve our vision of community under the limits of self-consciousness and physical scarcity which have been our terrestrial condition.

Even the best of our governments — democ-

racy — is based on the adversary mode which assumes that for one group to "win," another must lose. Cosmic community will stimulate the evolution of the self-governing process toward "wholocracy", as David Spangler calls it — government by attunement to a common plan and attraction to an unfolding evolutionary future which is sensed from within each person rather than imposed from outside by a coercive force, however benign.

Space communities will be designed as a synergistic whole from the beginning — the environment, the social relationships, the aesthetics. Each little new world will be a common creation. Our challenges, obviously, will be those of creators as well as creatures.

• • •

Economic systems will become wholistic — taking into account that we are all members of one body. But wholistic does not mean totalitarian or communistic, which are premature, coerced wholism. At the next phase of Conscious Evolution, with synergy, syntony and suprasexual attraction at work, we will be attracted into the whole for mutual enhancement. And the whole will be more biological than mechanistic, increasing the diversity and uniqueness of each of its parts. We will learn to play an I-win-You-win game rather than an I-win-You-lose game.

In space the little worlds will be built by human nature, in conscious cooperation with the processes of the Universe.

Basic needs of food, shelter, medical care will be met in community.

The rewards for excellence will not be more physical goods but increased opportunity for chosen action. Work will be experienced as that which we do to grow, to "self-actualize" as Maslow called it — a privilege not a burden. The highest status will be the freedom to take the creative steps, to explore, to evolve. Instead of this privilege being the prerogative of the few, it will be the opportunity of the many.

• • •

The evolution of consciousness from self to cosmic is already happening naturally on Earth. Imagine the stimulus to that natural development which comes from living for extended periods in cosmic communities in space, beyond the static and noise of the terrestrial world. The higher consciousness which has been "simulated" on Earth by disciplines of prayer, isolation, meditation, fasting, and drugs may be triggered more easily by weightlessness and the clean, uncontaminated environment of infinite space. The astronauts' experience suggests that a profound spiritual awareness arises naturally in the cosmos. If this is true, it will accelerate the evolution of cosmic

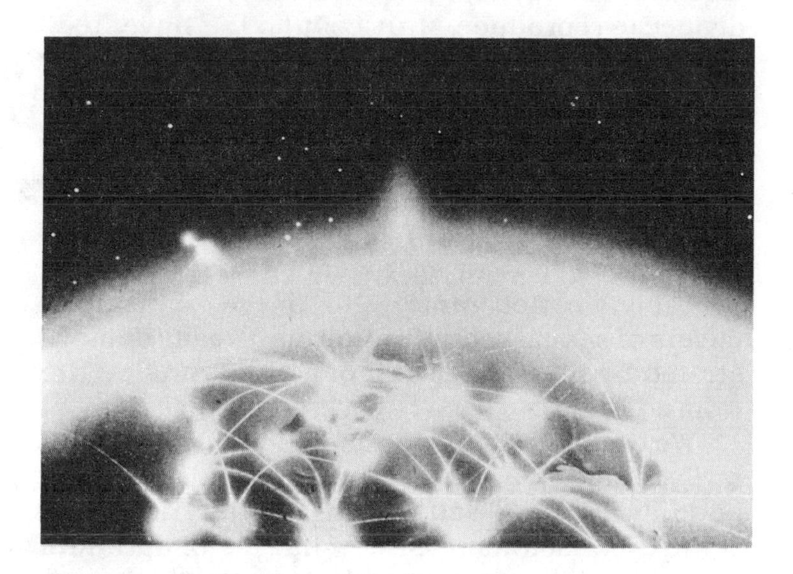

consciousness and community. Each one of these characteristics is mutually reinforcing, or synergistic.

COSMIC TIME

Cosmic time turns on as cosmic consciousness, community and action begin. At the dawn of human consciousness we sensed the limit of death and strove to overcome it. We never accepted the ultimate end. Reverence for life led to medicine, to control of many diseases and finally to molecular biology and genetics, the understanding of the language which directs the building of our bodies, and the code which programs us to live and to die. Gerontologists are studying aging as a disease, not an inevitability. We are attempting to reset the "clock of death," which was programmed millions of years ago with the advent of multicellular life and sexual reproduction — and is still ticking in us now, causing our cells to degenerate. Scheduled death was an historic design innovation, not an eternal verity. Single cells divided to reproduce. Multicellular life "invented" sexual reproduction and involuntary death of the individual. The result was the diversity of the species — newness — the improvement of bodies through recombining DNA — or information exchange through sexual exchange of genes!

In the next phase of evolution, Homa universalis has access to an extended life span (and probably, concurrently, to more conscious means of sexual reproduction, as already demonstrated by the conception of a child outside the womb.)

We are no longer programmed to die involuntarily. Through a whole spectum of abilities from body awareness, self-healing, better nutrition, cellular consciousness, new images of optimum wellness, deeper community, empathy, and attunement, coupled with understanding the language of the genetic code and our complex hormonal system, we gain the option of extended youthfulness.

• • •

From an evolutionary perspective, the meaning of extended life is extended action in the extended environment of the Universe. Increased longevity has no advantage to an Earth-only species on an overcrowded planet. It would be lethal. But for a universal species who will voyage beyond our solar system, it becomes an evolutionary necessity. The brief mammalian life cycle to which we now submit is much too short for cosmic life.

Genetics and astronautics arose in the same generation that recognized the limits to growth on the mother planet. We gained access to the language of our own genes — the building code of our bodies — at the same moment that we moved beyond the womb of Earth from which and for which these lovely mortal bodies were designed.

Those who choose to live and work in outer space will need the wisdom of the individual rather than a diversity of young new bodies. We may need to become "continuous humans" as Buckminster Fuller says in *It Came to Pass Not to Stay.*

> It took only two million years and
> Four and one-half billion human babies
> To establish a human survival beachhead
> Aboard the little
> Eight-thousand-mile-diameter
> Spherical Spaceship EARTH
> Whereby life could successfully realize
> Its highest known potential life span
> Possibly to continue indefinitely
> As one self-rejuvenating generation.
> Since all the vital parts
> Of human organisms
> Have now become interchangeable
> And many of them
> Have also become interchangeable
> With inanimate mechanical parts,
> And since human longevity
> Is continually increasing
> There is a good possibility
> That humanity is developing
> A continuous human

Who will persist in prime health
And youthful vigor
With the lessening of need
To replenish the population
With fresh baby starts.
The built-in drives to procreate
Will lessen and be manifest in a proclivity
Of females to camouflage as male
And males to camouflage as female
Thus suppressing the procreative urge
By superficial antipathetic illusions,
While permitting and promoting
Procreatively innocuous sex
companionships.

We are already evolving more thoughtful means of sexual reproduction along with the possible option of extended life. Sexual reproduction and death entered the evolutionary phase together, and they may pass out of the scene together.

• • •

We begin to identify, not with this particular bodily form we are now "in", but with the creative process which designs all new bodies. For we ourselves are becoming the designers of bodies — first to correct genetic defects and eventually to evolve bodies consciously designed for cosmic time and space.

New choices enter history. Death by choice. Life by choice. If death can be postponed, then death, like life becomes a choice. This is to be expected. We have learned from the fifteen billion year tradition that evolution is a freedom-raising experience. Unchosen, involuntary death is surely a tyrant. Disease is a dictator. It is obscene to imagine a universal human suffering from death by cancer or heart disease, like a helpless animal. The dignity of humans requires death, and life, by choice not coercion.

We see already not only the new science of gerontology, but also thanatology, the art of dying with grace, choosing not to extend one's life beyond the "natural" cycle (or what was natural for Homo sapiens during this phase of evolution.) Thanatology and gerontology. To die with dignity — perhaps to pass on towards life after death in a chosen manner, for those who sense the attraction of life in another dimension. Or to live on in the cosmos — life after Earth, consciously transforming this given mammalian body through total understanding of how it works and through building new environments, new worlds in space.

A new way of transcending — through body understanding and body building, rather than body failure and death — just as we always thought. Our religious premonition always assumed a generation would come that did not have to submit to physical death and would not be bound to this earth, as stated in the Bible:

> Behold, I show you a mystery
> We shall not all sleep, but we shall
> all be changed . . .
> For this corruptible must put on
> Incorruption, and this mortal must
> put on immortality, then shall be
> Brought to pass the saying that is
> Written, Death is swallowed up in
> Victory.
> O death, where is thy sting
> O grave, where is thy Victory?
>
> — 1. Cor. 15:51-55

COSMIC CONTACT

Cosmic contact is made as cosmic consciousness, action, community and time are established. Remember the Universe — the billions of galaxies, the multitudes of solar systems, some that may have life comparable to our own. In this epoch of our "birth" into the Universe, the eyes and ears of young Humanity are just beginning to open beyond the blue cocoon of Earth in which we have been gestating. The search for extra-terrestrial intelligence is on! The great listening dishes are attuned for signals. We are sending messages out into the vastness of space — not certain where to aim our signals. UFO research intensifies. Millions report sightings, and yearn for contact. We sense we are not alone and naturally seek to find others like ourselves. The starlight of curiosity awakens, fueled by the anxiety of meaninglessness, the fear of being isolated forever on a tiny planet in the corner of a small galaxy. We seek the larger community of which we sense we are a part.

Telepathy is studied as another option for communication in a universe where light travels too slowly for lively dialogue.

• • •

As we become young extra-terrestrials, capable of living, working and being responsible for ourselves in the Universe, it becomes increasingly likely that we will encounter other extraterrestrials. Like a new-born baby we are probably surrounded by life but are too immature to understand its language. We must integrate ourselves on Earth, develop our extended communications capabilities, learn the physical skills of operating in outer space — and open our telescopic and telepathic senses wide with intentional interest to discover our cosmic neighbors.

Now is a time of preparation. If a real visitation from another civilization were to be sighted by our planetary nervous system, the mass media, there might be planetary panic. We need to mature by developing our own cosmic consciousness, action and community to be able to respond to life that is more mature than we.

We expect the contact to be benign, based on watching our own development and witnessing the tradition of evolution, which is, at every quantum transformation, a leap in consciousness and freedom through synthesis of separate parts. The emergence of a more wholistic morality, an ecological ethic, is inherent in the transformation. Without it we will abort the process and become planetary dropouts. Misuse of power cannot escalate along with the powers of creation. There is a viability test for universal life. The very technologies necessary to achieve it are also the tools of planetary self-destruction. At the next stage we cannot survive if we are not "good."

To make the step toward universal life, we must mature morally, or we will destroy ourselves. The same is probably true for other planetary species. Those who make it past the viability test of high technology have no need or inclination to hurt anything or steal anything. They are capable of creating anything they need. They have become co-creators.

• • •

Throughout the ages we have been in contact with "voices," gods, signals, messengers, visions and visitors from outer space.

But these contacts almost always have come through to separate individuals. Only rarely has the experience been shared, as with the return of Christ in a new "glorified," imperishable body — and even then, only to a very few disciples. We have never had a shared planetary experience of higher life.

As we grow from self-consciousness to cosmic consciousness and become physically extraterrestrial, the likelihood of a common shared experience with other intelligent life will increase. Perhaps it will occur through our extended nervous system — our mass communication media. When this happens the separation of our Earth-bound life in the womb will be over.

As we unite on Earth and reach out into space, we may join a constellation of cultures — and the next loop, the next quantum transformation will begin.

COSMIC LABOR

Cosmic labor is a natural ability of a universal species. During the Earth-only phase of Homo sapiens, until very recently, our physical body was our main instrument of action. Now the body is becoming the house of thought, evolving from the instrument with which we build and move and defend ourselves. If we could take a picture of our real physical capacities now we would see our body has been extended into our machines — automobiles, bulldozers, telescopes, telephones, rockets, etc. Now these machines are being cybernated, automated and miniaturized. They are becoming intelligent, self-correcting. Our extended bodies work automatically to perform many of the basic survival tasks of our extended planetary body, as our own bodies are now automated to perform vital functions.

We do not have to say "Eyes see!" "Ears hear!" "Heart beat!" Our autonomic nervous system is monitoring and directing the hygienic functions to maintain our body so that our attention can be focused on creative tasks.

Now, throughout our world, cybernated machines, computers, space satellites and global communication systems will monitor automatically our planetary body by sensing devices surveying crop disease, weather, navigation, nuclear explosions. The production of many goods and services will be performed by automated machines. We will do by hand that which we choose to. As we enter our new cosmic home, the cosmic labor force will come into its own, providing absolutely vital intelligence and guidance systems for survival in outer space.

• • •

For an Earth-only species, cybernetics, like longevity, could be disasterous. It would take employment from growing populations and could be used for control of every aspect of our private lives. However, for a universal species, living throughout the solar system and beyond, cosmic labor will serve to emancipate the individual for the next set of creative functions.

We are moving toward the third industrial revolution — from slave labor, to people supervised machines, to the new cosmic Earth-space civilizations based on inexhaustible resources, cybernated machines and the liberation of the unique individual potential that is attracted to the process of evolution.

COSMIC INDIVIDUALITY

Cosmic individuality is the great gift to us personally of the next phase of evolution. The individual person, heretofore subjected to the tyranny of continual work to reproduce and survive, to early death and ever-present threat from the elements without and diseases within, will be free to discover what it really means to be human. We cannot judge Humanity yet. Most of us, most of the time have been enslaved to physical need. We are at our beginning.

Remember that we are still an unfinished species, not yet fully mature. Remember it is said we have only used five to ten percent of our mind-body potential — even the greatest geniuses. Remember that evolution is a consciousness-raising experience.

Now, imagine a world in which individual men, women and children are liberated from the past phase of creature human functions of maximum reproduction and survival tasks. Our daily survival needs for food and shelter are provided with minimum human effort. The productive capacity of a universal species, utilizing intelligent machines and renewable resources in an Earth-space environment, is astronomical.

Contra-ception becomes "pro-ception." Every child born is chosen, wanted and adored. Birth defects are a nightmare of the past.

• • •

Work means creative function. We begin to work in cooperative, synergistic organizations, in new communities on Earth and in space. Nature's wholistic tendency prevails again, through us. The untapped potential of individuals is stimulated by chosen vocation and participation in communities of share attraction, working freely together for the good of the whole.

Disease is overcome. We recognize that our health is our personal responsibility. Every thought, every act affects our whole system. We become masters of ourselves.

A new set of challenges then emerges — the challenge of a universal species at the beginning of its life beyond the womb of self-centeredness and planet-boundedness. We aim not at "utopia" which means "no place," but at life ever-evolving in a Universe which is itself evolving . . . an unimaginable future.

Does this set of possibilities seem idealistic? The answer is, yes, indeed! It created us out of stellar dust through ideas made manifest in form. (What is the DNA, the genetic building code of our bodies but an idea that builds form?) Evolution is continuing now, through us, to create new forms out of the old, according to the fifteen billion year tradition of transformation.

EVOLUTION IS A FREEDOM-RAISING

Freedom means the right to do your best, the opportunity to give your best to the future of Humanity. And freedom lives, experientially, at the growing edge where human creativity is challenged by the unknown. It is not a hot house flower.

"Freedom" in a closed system where everyone is cared for without challenge becomes a meaningless burden. Modern alienation, the "escape from freedom," is the painful heritage of a generation caught in the transition between the Old Age and New Age, between terrestrial and universal life.

In the coming period we are beckoned by a "choiceful future." Teilhard de Chardin said, "Union differentiates." As we integrate into one planetary culture on Earth and reach into the Universe for new life, an abundance of new, vital, evolutionary functions will open up to attract the full spectrum of human genius. At the moment we are in a narrow passage, where resources appear to be diminishing and people themselves do not feel needed. New work will draw forth untapped human creativity. The great over-endowed brain of Homo sapiens will be revealed as a vital reservoir of intelligence to transcend the animal condition and to become universal. The burst of genius, liberated from past limits, will make this planet a star of creativity!

Until we meet other life, we cannot judge our history. We have no other planetary culture to compare outselves to. Perhaps we have been a planet with a serious defect. Perhaps there are civilizations that do not have to endure the cruelty of a natural system in which millions of species eat each other alive to survive, and where individuals with sensitive nervous systems and love of life in their hearts have to submit to the pain of death, disease and the illusion of separation from the Creative Force of the Universe. Perhaps some planetary beings grew smoothly from animal to co-creator without the pain of the period of self-consciousness — or the "Fall." We do not know. But we do know that the future which we can preview is the pragmatic realization of the intimations of the saints and seers of all cultures.

• • •

Cosmic consciousness, action, time, contact, labor, individuality — these elements and others are the capacities needed to fulfill the inspirations we have received from the highest beings on Planet Earth.

Do we have the imagination and the will to live up to our new capacities? Of course we do. Because the evolution of consciousness, freedom, and beauty is a fifteen-billion-year tradition written in our minds and bodies. The intention to overcome limits through innovation and transformation is the most ancient tradition in the cosmos. That is why we have always known we would go beyond the mammalian condition to become cosmic beings. That is why we have never been satisfied by creature comforts. That is why we have always refused to live by bread alone and now refuse also to live with a scarcity of bread. Because it is only natural!

Our technologies are not ahead of our consciousness. They are barely catching up with our vision. Our technologies of transcendence are about to empower our vision of transcendence. We are about to act out in real terms in real time the dreams of the ages.

BIRTH

This vision of the future is a fulfillment of the intimations of the past.

We Are One Body — All People

We coordinate as a planetary body, attuning to each other and to the Designing Intelligence.

We Are In Contact With Other Life

The intimation of higher beings is affirmed. We have always been in contact intuitively. The esoteric or hidden is becoming exoteric and clear. Once we saw through a glass darkly. Now we see face to face.

We Are Immortal

We are not bound by the limits of this body.

We Are Universal

We are not bound by the limits of this planet.

We Are Other Life.
We are Higher Beings

Our innate sense of growth potential, our intimation of a higher state of being is true.

We are Conscious Co-Creators,
Partners With God

Mystical and secular, intuitive and rational forms of awareness unite in evolutionary consciousness, as we attune to the pattern in the process and assume some responsibility for the technologies of creation on a solar-system scale.

The future affirms the past. The root of the

word religion is *religare*, to bind back and make whole. We are reunited with our entire evolutionary past — from our cosmic conception through our birth into the Universe.

Know More Of The Creative Process Is The First Commandment Of The Universal Age.

We must know more of the laws of the Universe as we become more creative and powerful. The laws that guide us are the laws of evolutionary process and transformation. The precedent we draw from is fifteen billion years of success. As infinitesimal co-creators, we greet the infinite Universe with the humility of hope that we are loved, we are good, we are needed, we are capable.

We Reach The Tree Of Life

Through expanded love and expanded knowledge the hunger of Eve is fulfilled. The desire for deeper union with the Creative Intelligence is satisfied. The separation is over.

We Become The Second Couple

Adam and Eve were the first couple. They joined the masculine and feminine and together made a whole being, wherein they reached the Tree of the Knowledge of Good and Evil, separating from the animal world. To reach the Tree of Life, to have access to the powers of creation, each person must become whole, uniting the masculine and feminine, the yang and the yin, the rational and intuitive. Then we can unite whole being to whole being — co-creator to co-creator — at a higher level of intimate love, suprasexual and non-possessive, whose purpose is not only to reproduce biological bodies, but to co-create works for the future. The second couple reaches the second tree. Cosmic consciousness beings.

• • •

THE BIRTH MODEL

A Planetary "Birth" compared with a Biological Birth

At the time of **BIOLOGICAL BIRTH**	At the time of **PLANETARY BIRTH**
• Post-natal consciousness of each cell increases	Consciousness of each individual member of the body increases
• Capability for physical action	Capacity for physical action in "outer" space begins
• Coordination among cell members of the body deepens as body learns to function as a whole	Coordination among members of planetary body deepens
• Postnatal lifetime longer than prenatal life-span	Post terrestrial or cosmic life cycle longer than terrestrial life cycle
• Contact with "other life" — people!	Contact with "other life" — other universal species
• The physical body is automated and cybernated	Work of producing material goods and services is automated and cybernated
• Individual cells awaken to new functions	Individual person awakens to new functions

Without this higher form of intimate union, we are too uncertain in our loving powers to handle the powers of creation, and too unstable in cosmic consciousness to continue to re-member who we are.

This union of opposites within each person may be the most critical key to entrance to the universal phase of planetary evolution. A species whose men and women cannot learn to love on the intimate level, in a whole relationship without fear and possessiveness, will probably never gain the emotional maturity needed for universal life.

If this is an intrinsic viability criterion, it may be happening throughout the Universe. When we look up at the stars at night, and imagine the invisible planets as wombs for life, we can visualize other species arriving at the point of power, learning to love each other, whole being to whole being, and emerging from their planetary wombs to the universal world.

It is a loving Universe for those who are capable of becoming universal. Cosmic love, cosmic joy, the peace that passeth all understanding, the love that knows no fear or jealousy are the keys to the Kingdom of Universal Life.

And so the unfinished story of creation suggests new answers to the questions of the first universal generation:

What Is the Meaning of Our New Crises?
What Is the Purpose Of Our New Powers?

Those answers are:

- *The meaning of our new crisis is to activate our new capacities.*
- *The purpose of our new powers is Universal Life.*

. . . and we knew it all the time.

Earth Bound History Is Over
Universal History Has Begun

PART III

NEXT STEPS

Toward Creative Involvement

We have come to a psychological watershed. We need to act consciously and cooperatively for new possibilities which are unknown in their effects and unlimited in their potentials. No society — and especially individualistic democracies such as the United States — has much experience in conscious evolution, evolutionary management or long range goal-setting.

In the 1960s, people learned to organize against the injustices of the past. In the 1970s, we were stunned by the complexity of our problems and the newness of our condition. It is our challenge in the 1980s to learn to organize for the future. Most leaders are responding to problems with a pre-Copernican, closed world view. They interpret our problems as mistakes rather than stimulants toward innovation and transformation — a short-sightedness which could be deadly.

Time is not on our side; we cannot wait to activate our new capacities. Resources are dwindling. Costs are rising. Arms are escalating. Inflation is increasing. Human needs are growing. Population will increase for some time, no matter what we do. The situation is indeed serious.

We have a twenty or thirty year time frame to stop old destructive patterns such as exponential population growth, social injustice, violent means of conflict-resolution, misuse of non-renewable resources, environmental destruction, etc. We have the same time frame in which to initiate new patterns such as synergistic conflict-resolution, wholistic consciousness, life-enhancing institutions, renewable energy sources, and the development of a productive foothold in space.

How can we initiate a "politics of hope" and a "platform for the future" in time? We can no longer render unto Caesar what is Caesar's and unto God what is God's. The power of Caesar is too great. We must infuse temporal power not only with the love of all people as one, but also with the awareness that the whole human race has a future of unlimited possibilties in a Universe without end.

We must give blessings and moral sanction to our geniuses who are working at the growing edge of human ability. We must support those who are demonstrating the new capacities and endow them with resources to fulfill their gifts to the human species. As the wise men of old came bearing gifts to the Christ Child, our geniuses are coming to the cosmic child, Humanity, with gifts of life. They must not be rejected at this crucial hour of transformation.

We need a new kind of leadership from people of all races and backgrounds, people who are aware of the evolutionary potentials, attracted not to power for themselves, but to empowerment and enablement for all.

And in fact, such leadership is emerging.

Chapter 6

THE SOCIAL POTENTIAL MOVEMENT

Twenty years ago the human potential movement began with such seminal thinkers as Abraham H. Maslow, Victor Frankl, Robert Assogioli and others, who discovered, nurtured and affirmed the untapped growth potential in the individual person.

The next natural step is now beginning: a social potential movement that is identifying, communicating, and activating our collective capacities in all fields. This movement demonstrates what the human race can do together based on the reality of all our potentials —

spiritual, psychological, social and technological.

The founders of the social potential movement are such seminal evolutionary thinkers as Teilhard de Chardin, Sri Aurobindo, Buckminster Fuller, Arthur C. Clarke, Krafft Ehricke and Julian Huxley. They have identified the evolutionary pattern of Creation toward greater consciousness, freedom and order. They have seen that Humanity as a species is evolving into an interactive whole system far greater than the sum of its parts. It is this emphasis on the integral relationship of personal and collective evolution that distinguishes the social potential movement from all others.

This movement has as yet no accepted name. It is rising spontaneously in a wide variety of people who have caught sight of the reality of all our potentials and can therefore no longer accept false limits which could destroy our chances to evolve.

A key to the social potential movement is to imagine ourselves in the future as a more mature species. This is essential. For, as each individual needs positive images of the future to be fully well, so does society as a whole.

Recently we have had many images of social breakdown: Armegeddon, nuclear holocaust, biospheric collapse, social anarchy, etc. From the evolutionary perspective, we see, in contrast, a positive image of the future, a "*preview of coming attractions*," a view of ourselves as an ever-evolving species, ever closer to the pattern of creation, to conscious cooperation with God.

The ideas of the social potential movement are beyond present forms of conservatism and liberalism, beyond capitalism and communism. All these ideologies are based on pre-scientific technological capacities. They assume a world of inevitable scarcity and competition, in an Earth-bound environment where everyone is in a state of relative self-centered consciousness, working by the "sweat of the brow," living short lives. In other words, they are based on the present evolutionary state as the given condition.

It is obvious from the evolutionary perspective, however, that these are precisely the conditions which are being overcome through our emerging collective capacities. Clearly, we are not Earth-bound. Nor are we in an environment of scarcity, if you consider the whole Universe. We will not remain in a state of self-centered consciousness forever, for evolution is a consciousness-raising experience. We will not have to work by the sweat of our brow but rather by the instruments of our mind which can do the sub-human and non-human work. Our lives will not be short, but may be extended to an unknown degree through the wholistic application of medical advances and self-healing powers.

Evolutionary social goals are based on the activation of the full spectrum of our potentials, as a natural process, due not merely to human goodwill and planning alone, but also and more so, to the irresistible tendency in evolution towards greater consciousness, freedom and order.

The social potential movement is based on a new set of positive beliefs and values that nurture our transformation.

We can have wholistic consciousness.

We can live in a universal environment.

We can cooperate synergistically.

We can be liberated from unchosen work.

We can overcome disease and extend our lives.

It is natural that we transcend our present condition. For the nature of nature is to transcend limits.

The lessons of evolutionary change are applicable to our present condition. The patterns of transformation which created the Earth, life, multi-celled life and Humanity are still at work now, creating Universal Humanity out of young Homo sapiens.

LESSONS OF EVOLUTION

The social potential movement is also based on understanding and relating the lessons of evolution to our current crises. Remember the lessons that emerged from our study of the spiral:

Quantum transformations are traditional. Nature creates absolute newness through the synthesis of separate parts — cells out of molecules, animals out of cells, a planetary civilization out of billions of individuals. Newness comes from a more complex pattern, and we are forming a more complex pattern now. The force is with us. Newness is emerging.

Problems are evolutionary drivers. They stimulate potentials dormant in the system. Our problems are in fact driving us to new approaches in all fields — health, education, community-building, environment, energy, etc.

Design innovations with a technological substrate are at the basis of every quantum change. The genetic code, photosynthesis, language and symbols have led to quantum change. Now human-made technologies, from rockets and lasers to biofeedback machines, are in fact providing physical form for a quantum leap in capacity, evolving us from creature-humans to co-creative humans, assuming godlike tasks.

Wholism is inherent in the nature of reality. From hydrogen and helium atoms spewed across the Universe to the human being, nature's wholistic tendency can be seen in operation. And in fact, today we note a new level of integration as the world becomes one interacting ecological, economic, organic system, despite our institutional and cultural barriers.

Evolution is a consciousness and freedom raising experience. From molecule, to cell, to animal, to human, to us, stimulated world-wide by a culture-rich environment, nature has produced ever more conscious beings, ever freer to choose to evolve or die . . . until at last, our generation has arrived at the first age of Conscious Evolution. Our generation is the one selected to co-create the future of evolution on Earth.

Evolution proceeds by long periods of incremental change punctuated by sudden shifts, radical changes, like water turning to ice, or a bridge breaking, or a baby being born. We seem to be on the verge of such a sudden shift. The old patterns are breaking down. New patterns are forming under the surface.

At some point the new condition appears and the old is forever extinct.

From the evolutionary perspective this sudden shift is a radical transition from our Earth-bound, self-centered phase to our universal, whole-centered stage. This transition appears to be a normal planetary event, just what we would expect of an intelligent species who has gained its technological capacities and has almost reached the limits of its finite planet. It is as natural as birth. In fact, it is a birth! A planetary birth. Quite literally, we are being "born" into universal consciousness and action.

We have the capacity to survive and grow as a universal species. Both our problems and our potentials are driving us to innovation and transformation.

GOALS OF THE SOCIAL POTENTIAL MOVEMENT

The social goal of the future is the "self-actualizing society" — society operating at full capacity.

Just as each individual has a hierarchy of needs, so does society.

For the individual it is essential to satisfy basic needs for food, shelter, security and esteem; growth needs for chosen vocation and creative work; and transcendent needs for connectedness with a larger whole beyond oneself, as Abraham Maslow has pointed out.

For society, the hierarchy of needs is the same.

The first goal is to meet the basic needs of one hundred percent of Humanity through appropriate application of our full spectrum of abilities.

This goal is based on the assumption that society already has the technologies, resources, energy and intelligence to "live sustainingly at an unprecedentedly higher standard of living for all Earthians than has ever been experienced by any . . . through a design science initiative and technological revolution," as Buckminster Fuller put it. What we are lacking is not the capacities, but the initiative and revolution.

The goal of meeting the deficiency needs of all people is based on what G. Harry Stine calls the "third industrial revolution." The first revolution was agricultural — the domestication of plants and animals. The second was mechanical — the mass production of goods and services based on non-renewable resources. The third industrial revolution is "noological," coming from the human mind's understanding of the laws and processes of nature so as to create as nature creates, through innovation, synthesis and transformation.

The fulfillment of this goal requires a new concept of growth. Not industrial growth as we have known it, but new growth based on renewable resources, new sources of energy, new concepts of production, new accessibility of raw materials, new communities, new life styles, space colonization, life extension, and consciousness transformation.

At this stage of evolution, we recognize that we are in an indivisible Earth-space environment, and that we must develop that total environment as the material basis for the fulfillment of human potential.

We need a twenty year "Earth-Space Human Development Program" which aims at the restoration of the environment of Earth, the limitation of exponential growth here, and the involvement of Earth's people in a "creative revolution," not to redistribute existing wealth, but to co-create a new sufficiency through synergistic organization and proper use of new capacities.

As usual with evolutionary goals, this sounds overly optimistic, idealistic. It would be impossible unless you take into account the natural tendency of evolution toward synergy, wholism, and transcendence of limits.

This is the task of the social potential movement — to take the next step toward understanding our evolutionary potential and acting upon it.

To meet the basic needs of one hundred percent of Humanity it is necessary, for the first time in human history, to participate consciously in our own evolution, identifying and utilizing the full range of resources available.

To survive, we must evolve.

The second goal is to meet the growth needs of every person — the fulfillment of individual potential through creative vocation. In the context of the development of our Earth-space environment, in which basic needs are met by collective capacity and new frontiers are being explored, the untapped genius of the individual will be emancipated. During the early phase of our history we have been pinned down to repetitive tasks of reproduction and survival. As the full spectrum of technological and social capabilities become operative, we will be liberated from these tasks for new work, chosen work — the work of the next stage of evolution, building new worlds on this Earth and in space. This kind of evolutionary function will demand of each of us, not our bodies and brute strength, but our unique creativity. We see harbingers of this condition in pockets of relative affluence in the developed world, where masses of people are seeking self-fulfillment and meaningful function.

The key will be to connect individual creativity with specific social needs of an evolving society. Through the phenomenon of "networking" we are already witnessing the emergence of new vocations developed by shared interest and affinity. Networks for wholistic health, environmental restoration, consciousness expansion, cooperative entrepreneurship, human rights of all kinds; networks for space colonization, for life extension, for cybernetics, for the search for extraterrestrial intelligence; networks to overcome hunger, to create world peace, to unify the races, nations and religions . . . networks to improve everything under the sun are now rapidly growing beneath the surface of bureaucratic institutions, as Marilyn Ferguson writes in *The Aquarian Conspiracy*.

As individuals' basic needs are met with relative ease, they are freer to join these networks, which are decentralized and open to all, thereby discovering their chosen vocations, their evolutionary work beyond self-reproduction and self-maintenance toward self-actualization and life-enhancement.

We will move beyond the welfare state to the "creative state" in which people are free to do their best, and to give their best to the future of Humanity.

Obviously, we need a new education for conscious evolution, a metacurriculum for the future that provides every student with awareness of the macro-evolutionary history and access to all the techniques for personal and vocational development now available, largely outside academic institutions.

The third goal is to meet the transcendent needs of the human species. This means to consciously choose to overcome the past human condition itself. It means to evolve beyond our present limits through the activation of our full potentials. It means to overcome planet-boundedness, sub-human work, disease, and eventually, animal-like death. The transcendent need of society is to coordinate our collective capacities and thereby empower "ordinary" individuals with the capacity of the whole system. The need for transcendence is not met by individual quantum changes alone, but by collective and individual change. In the past, great individuals have transcended the limits of the human condition and performed miracles of psycho-physical transformation — healing, telepathy, attunement to higher intelligences, resurrection, etc. But the experience was not shared. Most people got it second and third hand through authorities and institutions which had not experienced it themselves. We could not experience these powers or perform these "super-human" acts ourselves.

However, as the collective capacities of society become activated, we witness the democratization of the transcendent. It is happening already. Everyone can see around the world with the speed of light via television. Anyone can leave the Earth in a flaming chariot, the space shuttle. All of us can be put to sleep and awaken with a new kidney or a new heart, as medical science normalizes miracles and gives them as an inheritance to every child born.

To use an evolutionary analogy: today each child inherits the biosphere, the living layer of life, and the noosphere, the thinking layer of cultural heritage of the past — the works of millions of years of evolution.

Now, imagine that only a few generations in the future, every child will also be given a cosmic civilization as its rightful inheritance. In order to reach this point and to meet the transcendent needs of society we must make collective, evolutionary decisions to allocate resources to transcend current human limits — in genetic engineering, in longevity, in intelligent machines, in space industrialization. These decisions cannot, yet, be made by small groups alone. They require of us new social functions.

It is obvious that the fulfillment of the goals of the social potential movement: to meet basic, growth and transcendent needs of human society, cannot be realized by our current institutions. We have the capacities. We lack the institutions.

EVOLUTIONARY DECISION-MAKING: A NEW SOCIAL FUNCTION

No existing institution is capable of making such choices on its own. Neither the corporation, nor the state, nor the school, nor the church has authoritative guidelines for the transformation. Society lacks mechanisms and processes for long-range, evolutionary decision-making.

What is required, as usual, is a new synthesis,

not of institutions, but of functions, in a pragmatic way, yet to be devised.

The synthesis will bring together science, art, politics and religion in some form yet to evolve.

Science is our capacity to understand the processes of creation and through its offspring, technology, to enter the process consciously in order to cause change.

Politics is our collective capacity to act, to allocate resources, to accomplish that which cannot yet be achieved by small groups and individuals.

Art is our capacity to envision and imagine that which can be, but is not yet . . . an increasingly important function as we gain the ability to alter our bodies through genetic engineering and build whole new worlds in space. What we imagine, we can become . . . increasingly.

Religion is our capacity to know the intention of Creation, the will of God, by direct experience, by inner contact, attunement, meditation and prayer.

Science provides the knowledge, politics the collective will, art the shared envisioning, and religion the values and directions based on internal knowing.

We need a new "ground of the whole," synergistic environments like the Syncons which offer a non-threatening opportunity for individuals to meet according to their functions, to listen to each others' long range goals, to match needs and resources, and to make collective decisions based on the full spectrum of capacities of the whole body politic, in an all-win mode.

Such experiments in group decision-making by consensus, attraction and attunement are occuring all throughout the world in new communities such as the Findhorn Community in Scotland, the Institute for the Study of Conscious

Evolution in the United States and Auroville in India. These groups and many others in all parts of the world hold evolutionary values that recognize our profound potentials, offer educational programs that empower others to express their life purpose, and model for the world new modes of living and being. They are all breaking ground, each in a unique way, for the emergence of a new civilization that is in creative harmony with the Universe.

SYNOCRACY: THE EVOLUTION OF SELF-GOVERNMENT

Are we yet willing to move from democracy toward "synocracy," that is, toward government by synergy, in which individuals participate freely in the whole community through creative function that is mutually beneficial to the whole and to the individual? Will we evolve toward a new form of self-government by self-organization and alignment with the patterns in evolution as a shared experience? We now need government by attraction, rather than coercion, a non-coerced wholism in which the freedom of the individual is enhanced, not surpressed, by union in the whole.

Is not a new synthesis of the functions of science, politics, art and religion, a natural evolution of human self-organization that follows nature's tendency to form whole systems?

It seems logical. We certainly cannot assume that a co-creative, universal species will continue to remain in divided and divisive relationships, governing itself by coercive systems, no matter how benign. For the power we have inherited (such as nuclear energy and bio-engineering) cannot be contained by ordinary laws. There is no police force on Earth capable of protecting us from ourselves as we learn how the atom, the gene and the brain work. A single crazy individual can bomb the world. One angry person can poison a city's water supply, and this is but the beginning.

Every great religion has forseen the reality of the internatization of the law through voluntary alignment of the individual with the patterns of Creation, or the will of God. Are the mystics, saints and visionaries, those in "cosmic consciousness," evolutionary throw-forwards? Precursors of the new norm? I think the answer is, yes.

As we evolve toward our universal phase, we will either all become capable of such a free, non-coerced alignment with the wholistic tendencies of nature, or we will self-destruct. Just as we can not give a car to a two-year-old child to drive, the powers of co-creation and Conscious Evolution are too great for a self-centered species.

The issue really seems to be: Evolve, Humanity, or die! Remember, "We shall not all sleep, but we shall all be changed," As Paul of Tarsus said almost 2000 years ago.

NEO-NATURAL SELECTION

We must assume that self-centered consciousness is only a stage, and that some evolutionary process is at work, beyond human design, selecting the whole-centered from the self-centered, empowering the former and disempowering the latter. In fact, evolution could be defined as a process of selection of ever more conscious individuals.

Is the human race composed of a variety of psychological species, forming a Bell Curve from extreme self-centered to extreme whole-centered?

As the social environment changes to greater interdependence and interconnectedness on the planet, and as each individual gains greater access to destructive and constructive power, a natural selection process may be occuring. We may be selected according to our psychological state of being. Those in a state of self-centered consciousness, who resist change and seek security in the past, or through the suppression of others, are even now increasingly under stress, going against the grain of the times, which is toward a more complex whole system. On the other hand, those who are attracted to participate in change and who are desirous of closer empathetic function with others, are already animated by the vigor of their inner drive to create, and are flowing with the patterns of creation.

Perhaps, as we evolve through this period of transition, we will emerge as a whole-centered species, that is, everyone who exists will naturally experience themselves as part of the same Creation, animated by the same creative drive to realize a unique function for the good of the part and the whole. Self-centered behavior patterns will have proven to be non-adaptive, non-viable, and will become extinct.

Some such selection process must have occurred at the beginning of self-consciousness. Could it be that a similar process of selection is at work among us now at the dawn of whole-centered consciousness, selecting for those psychological types capable of participation in the whole? Will self-centered behavior become extinct?

Again this seems logical. After all, every stage of consciousness is but a passing phase. At one point there was no self-consciousness at all! It must have seemed extraordinary and unstable in the animal world, just as wholistic consciousness is unstable and unusual now. However, at some point, self-consciousness became the new norm and no pre-humanoids were left.

It may follow, that at some point in the future, wholistic, evolutionary, co-creative consciousness will become the new norm, and self-centeredness will be as extinct as pre-humanoids are now.

Thus, at its growing edge, the social potential movement envisions a "eupsychian society," a society composed of self-actualizing people, people who are living up to their potential, people whose basic needs are met, who have chosen work, and who are transcending the limits of self-centeredness and the creature-human condition, people who are living in a cosmic civilization which has begun to act out the ancient visions of Paradise, the New Jerusalem . . . a new heaven and a new Earth. The metaphysical becomes the actual. The eternal ecstatic visions of "heaven" become the next stage of evolution as we end this phase of life and embark upon the new.

The new partnership between creature and Creator becomes a pragmatic necessity rather than a visionary dream.

THE VISION THAT DREW THE PIONEERS ACROSS THE SEAS

There is a special role for the descendents'of the pioneers that founded the United States of America. We are the children of the discontent, the risk takers, those who would not put up with something. The freedom lovers of the whole world came here to try something new. The United States is not a nation, it is the people from all over the world who dream of something better.

Our forefathers were inspired by ancient wisdom which is written on our dollar bill. It states the evolutionary goal of the world in clear terms.

E Pluribus Unum — out of many one. Unity in diversity. Synergistic democracy, "synocracy," wherein every person freely cooperates in the evolving whole, connected by creative vocation, fulfilling self-development and the development of the whole in the same act.

Novus Ordo Seclorum — a new order of the ages symbolized by the unfinished pyramid and the cosmic eye. From an evolutionary perspective this means that the great building power of the world must be connected with the spiritual wisdom of the world, to build a new world on this Earth and many new worlds in the Universe, "a new heaven and a new Earth," in alignment with the intention of Creation, which is forever higher consciousness, freedom and order.

Annuit Coiptis — He favors this enterprise. "God," or the pattern of Creation, favors the evolution of a universal society in which each person freely cooperates to build a new order.

New worlds on Earth, new worlds in space, a choiceful future for all. This is the appropriate evolutionary goal of the social potential movement. It is the proper next step for the peoples of the United States and the world, to build together an infinite future in which every woman, man, and child of Earth will be free to do their best.

This vision of our future is not new. It is older than the world, older than time. It is encoded in our genes which hold the memory of the original act of Creation. It is written in the sacred scriptures of the world — Nirvana, Paradise, the New Jerusalem. It is inscribed in the political systems of people around the world: "all men are created equal;" liberty, equality, fraternity;" "from each according to his ability, to each according to his need."

And now, by good fortune, we in this generation, have the chance to act consciously on the vision of the past. For ours is the power, ours is the crisis, ours is the choice as to whether the entire history of life on Earth is fulfilled, or destroyed.

Not only would it be difficult, it would be impossible if we required to accomplish this transition from Earth-only to universal life by human planning alone. The situation is too complex, beyond the modeling capacities of our design systems . . . so far.

But remember, we are part of the magnificent evolutionary processes that created us and that are still creating us now. We are not doing this alone. We are doing it in an evolving Universe whose processes and laws are operating through us, whether we know it or not.

This is the basis of evolutionary faith: *The Designing Intelligence that organized the Universe, is still at work in us now.*

And remember that evolution proceeds by quantum leaps, sudden shifts, vastly accelerated events in time. Could it be that we are on the threshold of such a quantum leap, that will radically and forever alter the world.

It is predicted in a negative way by environmentalists and anti-nuclear forces who speak convincingly of the biospheric collapse and planetary self-destruction if we do not change our behavior. It is prophesied by the fundamentalists who quote scripture about the sudden end of the world . . . the imminence of Armegeddon, an event perpetrated by higher beings for the purpose of cleansing the Earth of selfish behavior.

It is also foreseen by evolutionaries, the social potentialists who see a sudden shift wherein the innovating networks of shared attraction in all fields — health, space, human potential, environment, consciousness expansion, etc. — will rapidly interconnect and help the system self-transcend as the network and the bureacracies fail . . . like a kaleidoscope suddenly changing its pattern.

THE PLANETARY SMILE

I have envisioned a positive sudden shift . . . a planetary smile.

When a baby is just born, it does not know it. It feels the pain, and cries, until its little nervous system connects . . . then, it stops crying, opens ite eyes . . . and smiles.

We on Planet Earth are a new-born planetary species. We have been born into the Universe, and we are almost totally inter-connected by our mass communication media, the nervous system of the planetary body. We are feeling the pain of the whole body, and we are crying.

Could it be that when our nervous system connects, and enough of us focus together on one thought for one shared instant in time, we shall all smile together . . . a planetary smile, billions of people at once connecting. It would change the world forever.

I have written a description of a planetary smile in my Evolutionary Journal. This is a form of meditative writing which lets deep intuition express itself without judgement. Whether or not such an event will happen is not known. But it is possible. And if it were to happen, we might avoid the destructive process of selection toward which we are now heading.

(Note this style of writing. When you begin your own Journal, you might try such meditative envisionment. Can you imagine an event which could change the world for the good all at once?)

EXCERPT FROM AN EVOLUTIONARY JOURNAL

"There are x number of human beings on Planet Earth now in whom the flame of expectation for something better burns. It is enough.

This flame can be raised in millions at once by the shared experience of the reality of our potentials for life as Universal Humanity, life as co-creators, co-operating consciously with each other and with the patterns in the process of evolution (known to deists as God and to communists as the dialectic.)

At some point, a series of planetary events could occur, wherein people on Earth are linked with people in outer space (on the space shuttle.) They are naturally joined in a common thought pattern which affirms the flame of expectation in each individual. (Just as they were linked during the lunar landing by a common experience of an Earthian setting foot on a new world.)

A new psychological field of force is generated by the space-Earth empathetic mind link, wherein each person is thinking the same thought at the same time . . . We are one. We are whole. We are capable of a universal future.

Each individual who is consciously connected to the field, and many more who begin on the periphery and get attracted, has an experience of profound inner excitment, anticipation and empathy.

The mass media is amplifying pictures of people experiencing the same feeling of oneness as Earth People, and the same expectation of a cooperative life beyond the tribulations of conflict, hunger, disease and war.

At a certain point, the amplification of the empathetic thought pattern goes exponential, doubling every few minutes *on a planetary scale.*

People on the street, in shops, in homes in hospitals, in schools, in cars, in planes, in conferences, in the halls of government, industry, labor, and academia, know something is happening world-wide. As the word of assassinations spread very quickly, so could the word of the planetary smile spread.

Planet Earth, like any other body is composed

of atoms in action. When the atoms are oriented in the same direction on the physical plane, magnetism occurs. When the atoms are oriented in the same direction on the psychological plane by shared thought, psychomagnetism occurs. We have seen the phenomenon many times in parades, concerts, and celebrations.

The planetary smile turns on the power of mass attraction for the good. It intervenes in the sick decline toward global confrontation and mass destruction.

As the psychomagnetism builds between people on Earth, between people and the Earth itself, and with people in space, *the Earth may become receptive to signals from Universal Intelligence as a shared experience.* What has occurred over the millenia to gifted individuals, could, due to the boost of mass attraction amplified by mass media, become available to *millions or billions of people at once!*

The shared experience of higher intelligence, loving presences, higher selves, evolved beings, is a real possibility. For we ourselves are becoming evolved beings. We ourselves are becoming higher life. We may not be alone in the Universe. We may be surrounded by life, intelligence, and love.

That life has been experienced in diverse ways by different cultures. Inner voices, visions, commandments from a legendary line of "gods" have lifted humans from the animal condition to the expectation of life ever-evolving.

The Judeo-Christian world has experienced the loving presence personally as the Christ. They have expected a "second coming." It has not happened yet. Other people have other similar expectations, which are also unfulfilled. None has fully known what the details of that expectation are.

These global expectations may be accurate racial knowledge that evolution does not proceed by incremental steps but by quantum leaps, in the fullness of time, which no one can predict. *From an evolutionary perspective, the fullness of time is now.*

A PLANETARY PENTECOST?

The experience known as "the pentecost" startled the disciples three weeks after Jesus' death. They were in a small room . . .

> "They were all with one accord in one place. And suddenly there came a sound of heaven as of a rushing mighty wind, and it filled all the house where they were sitting. And there appeared unto them cloven tongues like as of fire, and it sat upon each of them. And they were all filled with the Holy Ghost, and began to speak with other tongues, as the Spirit gave them utterance. And there were dwelling at Jerusalem Jews, devout men, out of every nation under heaven. Now when this was noticed abroad, the multitude came together, and were confounded, because that every man heard them speak in his own language. And they were all amazed and marveled, saying one to another, Behold, are not all these which speak Galilaeans?
>
> And now hear we every man in our own tongue, wherein we were born?
>
> And they were all amazed, and were in doubt, saying one to another, what meaneth this?
>
> Other mocking said, these men are full of new wine.
>
> But Peter, standing up with the eleven, lifted up his voice and said unto them, Ye

> men of Judea, and all ye that dwell at Jerusalem be this known unto you, and harken to my words.
>
> For these are not drunken, as ye suppose, seeing it is but the third hour of the day.
>
> But this is that which was spoken by the prophet Joel.
>
> And it shall come to pass in the last days, saith God, I will pour out of my Spirit upon all flesh: and your sons and your daughters shall prophesy, and your young men shall see visions, and your old men shall dream dreams."
>
> Acts 2:1-17

Is this going to happen to us?

The illusion of separation would be overcome for one golden instant of time. Love would replace fear. Hope would replace despair, on a planetary scale, allowing the positive to be reinforced everywhere. The effect of such events are unpredictible.

The Planetary Smile, the Planetary Pentecost, the Great Reunion, the Second Coming, the shared Cosmic Birth Experience, the Contact with Higher Life are *real possibilities*, as real, perhaps more real than the possibility of nuclear holocaust. After all, we are the product of fifteen billion years of crisis and transformation. Perhaps we are living through yet another natural creative change.

We have examined in current times, with great perception, the possibility of unprecedented breakdown. But we have not examined with equal care the possibility of unprecedented breakthrough.

We have not asked, for example, such questions as these with real seriousness:

How does the whole system communicate to its parts? Is there a planetary "DNA," an encoded plan for planetary conception, birth, universal development? Are we the RNA messengers of a universal template, signaled to build according to inner motivational signals?

What form of energy is used to attract particles to form whole systems? Is it operating now? Is the affinity we feel to certain people and tasks the experience of a planetary body's members being pulled together to form social organs?

What is God's communication system? What are angels, really? Are they evolved beings? What does the Bible mean when it speaks of a "sound" being heard by the "elect" of all nations? Are such sounds something like extremely low frequency waves generated by the power of mind-at-a-distance amplified by electronics — a noological communication system, mind to mind, from more evolved beings to less evolved beings? We can already perform such communication feats through psychotronics. Why couldn't others do the same?

Is there a "global interference pattern," a "four dimensional dynamic hologram." Is everything that's ever happened, going back billions of years in the planet's history, still encoded now because the effect of those waves has only attenuated, never died out? Can we access this global interference pattern, decode it, develop a model of its contents, and utilize the model to access anything we want to know that has happened in the past? Remember "syntony," the increased capacity of the individual to resonate with the patterns of creation. "Perhaps in syntony with everything that exists, we can access that global interference pattern and use it, not just as a source of information, but as a way to communicate. Are those waves passing through our biological sensors right now? Are we all contributing to that global interference pattern as we read these words? Can we access it consciously?" as Lawrence de Bivort speculates.

Is the esoteric concept of an "akashic record" a real universal computerized data bank? Are the myths and legends of our ancestors the early racial intuitions of the next stage of evolution, which we are in fact, acting out now?

How would we look to pre-humanoids? Wouldn't we look like gods, like supernatural beings, when we are merely beings at the next stage of evolution?

No one knows whether such possibilities are fact or fantasy. They have not been investigated in a verifiable way. Yet millions of us feel as though we are on the threshold of a breakthrough in consciousness. Now we see through a glass darkly. Soon we will see face to face. The transcendent dimension of life hovers tangibly close, not because we are more spiritually gifted than our forefathers and mothers, but because the whole planetary system is linking up; the noosphere, the thinking layer is thickening. The hour of the unexpected is at hand.

As we live our daily lives, taking small steps to improve our health, to reunite our communities, to solve our problems one by one, let's remember that we are doing these individual acts in a larger system, which is itself shifting toward an interconnected whole, which will affect every part, over and beyond anything we do as individuals (just as cells in the body of a baby are all affected by the birth, no matter what their specific function.)

The point is to open our minds to the reality of evolutionary transformation, to stop seeing the world as a series of still snapshots and start seeing it as a multi-billion year process of creation going on now, through us, as nature's growing tip of change on Planet Earth.

THE PURPOSE OF EVOLUTION

This context transforms the content of our lives. Scale creates phenomenon. Take the largest possible scale — the universal, evolutionary scale. Stand beyond the whole system, seeing for one flash of time as the Creator might see. Witness ourselves as co-creators. Each act we do is an expression of the whole process of Creation, linked subtly and simultaneously with acts everywhere throughout the Universe, which is everywhere alive with the same intention we experience — the intention to evolve to higher consciousness, freedom and order. Now, the Universe comes alive with life, all matter is transmuted to consciousness.

From the perspective of a co-creator, we see the Universe as a nurturing ground for godlings. We see evolution as the mechanism for the selection of ever more conscious beings. We see history as the discovery of how to be godly.

As we take small steps toward self-development and the betterment of the world, we place those steps in the context of the evolution of the Universe. We see ourselves aligned with all being everywhere. And even deeper, we reach beneath the grand drama of evolution for the eternal aspect of reality — the total potential of the whole cosmic system, that which has been, is now and always will be. Through meditation, attunement, prayer, and inspiration we connect with the eternal. As we lean into the future casting our weight upon the waves of change, we seek humbly to be guided by the Divine intention that has created, is now creating, and will forever be creating the Universe and everything in it.

Chapter 7

THE EVOLUTIONARY JOURNAL

This chapter includes a series of ideas which emerge from the evolutionary spiral. Many of them represent lifetimes of study and the exploration of countless individuals. While different ideas will be of significance to different people, each idea is an integral part of the whole.

SUGGESTIONS FOR USE OF JOURNAL

Read each idea at a time most convenient for you, perhaps focusing on one idea for a week at a time. After thinking about it, begin writing in your journal. (A loose leaf notebook works best.)

Let your thoughts flow as spontaneously as possible. I have found that combining journal-writing with meditation can work very well. I just sit in a comfortable chair, close my eyes and begin relaxing. To achieve a more relaxed state, some people visualize themselves at the beach or in a serene outdoor setting. Others contemplate God or simply observe themselves breathing in and out. I started learning to meditate by repeating "The Lord's Prayer" until I could concentrate without my mind wandering through the whole prayer.

Once you have relaxed, focus on the idea for the week, and just observe and record whatever comes into your mind. As you allow the thoughts and images to surface without judging how "good" or rational or realistic they are, you will allow more and more information to flow from your own "inner knowingness."

If you have never kept a journal, this is a wonderful time to start. Your weekly writings may not seem significant to you at first, but you will be amazed when you look back at what you have done — at your insights and growth as the weeks go by. It is not necessary to follow the instructions precisely. You may find yourself writing poetry, drama or music, drawing pictures, dancing or doing something utterly unexpected. While the field of "Evolutionary Arts and Letters" is very new, it is definitely unlimited. New forms are needed. Be experimental and share what you do with others if you care to. The results of this kind of exchange are unpredictable.

Once you have completed this series of 13 ideas, (enough for a quarter of the year if you focus on one a week), we suggest you create a list of 13 additional ones, culled from what comes up in your journal-writing, perhaps working with someone else who is keeping an Evolutionary Journal. As you progress, you may find that more information is surfacing within you to consider and study. In fact, at some point you may decide to use your own Journal as a jumping-off point for either conducting an on-going study group or actually initiating a class in futures studies at a local community center or college.

Remember, just as this is your Journal, the future we are discussing is yours, too, to make of it what you will. One way you can make it the positive, fulfilling one that this Journal indicates is by sharing these ideas with more and more people who are open to them, thus igniting that spark in them, as one has been ignited in you.

In the Appendix you will find an Evolutionary Resources Section. All books referred to in the text are listed in the library. As you read these books, take notes, if you like, as part of your Journal.

We also suggest you start a Book of Connections — your list of people and activities that share the attraction for the future. This can become a invaluable networking tool for you and others who share your commitment to the future.

Now lets begin working with the Evolutionary Journal.

IDEA

YOU ARE A MICROCOSM OF THE WHOLE PROCESS, THE COSMIC EVOLUTIONARY SPIRAL AND YOUR PERSONAL STORY ARE RELATED, AS ABOVE . . . SO BELOW.

In "The Theatre for the Future" we "re-member" our cosmic past. We take the universal perspective, step outside our planet and imagine ourselves slightly in the future — a more mature species, less than one hundred years hence. We collapse time, compress eons into seconds, to view our past as a "photogenesis," a rapid time-lapse film sequence, a movie of creation from the beginning of the Universe to the present and beyond — to our species in the future.

Now, you can relive your own life as an evolutionary spiral, applying to it lessons learned from our cosmic past. We are going to focus on a "local event" — a tiny yet vital part of the whole evolving Universe: yourself.

THIS WEEK

First relive the Cosmic Evolutionary Spiral. You body and mind are the living product of the Creation. The memory is alive in you. Try to awaken it. Feel the explosion, the dispersion of gases, the condensing into solid Earth, the origin of life, the rise of plants, animals, the dawn of self-consciousness, the present crises — and our future with all potentials actualized and operating harmoniously. Dance it; here the music of Creation; see the images; sense the Designing Intelligence within you at work, in the beginning, in the present and in the future. Write in your journal any thoughts that occur to you.

Now relive your own personal evolutionary spiral, the story of your life. You may find it useful to draw a spiral in your journal, dating the points of crisis, the nature of the "evolutionary drivers," your "design innovations" (how you responded to the crisis), and the quantum transformations (how you differed after the crisis: what was new, different?)

• • •

Now look at your present situation. What are your problems? What strengths do you have to overcome them? What innovations are possible? The nature of your response to crisis determines your future. Do you choose to adapt to limitations, or to innovate and transform?

Imagine yourself in the future; what it will be like when you mature and use your capacities harmoniously.

To help yourself, think of a "compass of joy." Identify on your spiral your past moments of joy. Go back and relive them. Describe them. See if you can visualize any "light" out front in your future. Is there a new joy coming? What is it like? Try to experience that coming joy. Describe it in your journal.

What steps are you taking in your life to realize that attraction?

Above all, do not judge yourself. We are all in a state of becoming. Judgment and guilt can destroy our natural growth.

SUGGESTED READING:

• Hubbard, Barbara Marx. *The Hunger of Eve — A Woman's Odyssey Toward the Future.* Stackpole, Harrisburg, PA 1976 (Available From Evolutionary Press)

This a personal story of evolutionary struggle, innovation, and transformation — which is still continuing! It is an example of how the personal and planetary fuse, and may be of help in reliving your own past.

• Ferguson, Marilyn. *The Aquarian Conspiracy*. J. P. Tarcher, Los Angeles, CA 1980

An excellent description of the personal/ planetary transformation — with one major exception — Marilyn focused on inner growth alone. She does not balance the inner with the exploration of outer space and the development of transforming technologies in astronautics, cybernetics, genetics, etc. The evolutionary perspective provides a synthesis of inner and outer development.

• Emerson, Ralph Waldo. *Essays*. Especially "Self-Reliance."

One of the founders of transcendentalism, a pillar of current transformational thought.

IDEA

OUR PERSPECTIVE IS THE UNIVERSE.

There are billions of galaxies, vast clusters of stars gravitationally attracted into structures and scattered throughout the remotest depths of the cosmos. Our most sensitive astronomical instruments can "see" more than one hundred million galaxies, each with billions of stars! The total number of galaxies in the Universe is far greater. New generations of space telescopes will bring many more galaxies into our ken.

Only in our generation did astronomers begin to grasp the awesome extent of our Universe. The young eyes of science, the extended sensory system of humankind, are just beginning to see where we are, where we came from and where we may be going.

Most people associate "nature" with life on this Earth. But, in fact nature is the whole Universe. At night we catch a glimpse of our natural environment outside this little blue cocoon in which Earth-life has been developing for the past three and a half billion years.

Nature does not only exist in the present. It is also the past which lives in us now, and the future which is potential in us, like a seed.

THIS WEEK

The environmental movement has focused on a definition of nature as the existing balance of all life within our biosphere, an environment which is being threatened by crisis — evolutionary drivers such as pollution, the extinction of endangered species, overpopulation, etc. Taking into account our fifteen billion year past, our step into the Universe, and the immensity of the Universe itself, write your own definition of nature.

• • •

Imagine yourself as you now are on Earth. Draw away from yourself outward, until the whole Earth comes into view. Draw further outward until your galaxy appears as a whole — now further and further to the edge of the expanding Universe.

Be aware that everything that occurs throughout the whole system is related to you, and whatever you think or do, is a vital part of the whole. Write in your Journal any thoughts that occur to you.

SUGGESTED READING:

• Jastrow, Robert. *Until the Sun Dies.* W.W. Norton, New York. 1977

This is the best scientific description of the Creation. Jastrow confronts, tentatively, the mystery of the Creator — and backs away, leaving the subject open to evolutionary seekers, who fuse the secular and the spiritual forms of knowing.

• Sauber, William. *The Fourth Kingdom.* Aquari Corp., Box 1966, Midland, MI 48640, 1975

A fascinating theory of evolution which sees Earth developing its "seed pods" to reproduce Earth life through space arks . . . "for one day the sun will expand and destroy our mother planet." He sees technology, the "fourth kingdom," as natural as the mineral, vegetable and animal kingdoms.

• McWaters, Barry. *Conscious Evolution.* Evolutionary Press, San Francisco, 1982

The first basic text on the vision and principles of Conscious Evolution. McWaters bridges the gap between the metaphysical and scientific perspectives by offering a multidimensional view of our capacity to co-create a positive future. Essential reading.

IDEA

YOU ARE NEEDED FOR THE EVOLUTION OF THE WORLD.

You have a unique vocation in the whole, a next step which is calling you from within to extend your capacity and involve yourself in the world by doing that which you sense to be your best. The better you identify your unique function, the more deeply you can integrate into the whole.

In the next phase of evolution, our vocation and our "job" will increasingly merge, so that we will "earn our living" by living out our chosen function — that activity which gives us joy and through which we give our creative best to others.

Your chosen function may be an expansion of what you are already doing, or something new.

If you are not responding to your calling you are probably feeling deep frustration. Don't suppress it. Cherish it. Listen to it. Act upon it. The frustration is a divine discontent signalling you to respond to some vital growth need in yourself and in your planet.

Remember the idea of "syntony". We are assuming that the evolutionary process is intensifying its motivational signals at this time of quantum transformation. It is as if we are gaining more direct personal contact with an unfolding plan — a plan that includes the organic growth tendencies of a planet at this phase of development. Your vocation is a vital part of that plan and is encoded in your unique set of capabilities.

THIS WEEK

Have confidence that you are an integral part of evolution and that your personal frustration is a planetary growth signal guiding you toward your next step which is an aspect of the evolution of the world. You are not progressing for yourself alone. Everyone grows, as each of us learns how to take our own next step.

• • •

Take time to experience your growth-frustration, your signals. What is it telling you about your next step? Write down any ideas you have in your journal. Do not criticize yourself or edit your thoughts. The analytical intellect will obstruct your intuition if it pounces on every tentative new possibility to discover what might be wrong with it. Your intention is to nurture the green shoots of growing function in yourself. Learn to be non-judgmental, tender, kind and even parental to yourself, as you would to a dear child. At our growing edge, we are continually fresh, sensitive and new.

• • •

Let your mind wander freely. Visualize yourself doing what you most love to do. Imagine ways to continue doing it, expanding it, building on it. The law of attraction will guide you to the next step. At this phase of evolution, we are proceeding less from necessity and more from attraction. Shift your focus from your lacks to your loves and follow the compass of joy.

• • •

Remember your "peak" experiences, those moments of total well-being, unity, relatedness, sense of being "in the flow" when you and the process are one. Describe those experiences in your Journal. Identify common elements. What were you doing? Where were you? Who were you with? These peak experiences are guides to the next step of vocation. The aim of an evolutionary personality is to live at a new norm of creative involvement in the world, to escape from the "self-concentration camp" and be free to be totally committed to what you most love to do.

SUGGESTED READING

• Maslow, Abraham H. *Toward a Psychology of Being*. D. Van Vostrand Co., N.Y. 1968

Maslow was one of the fathers of humanistic psychology — which triggered the human potential movement. He studied well people, rather than the sick. His purpose was to help people overcome self-imposed limits, to become self-actualizing, that is, ever-evolving. A key and seminal work.

• Assagioli, Roberto. *Psychosynthesis — A Collection of Basic Writings*. Viking Compass Book, N.Y. 1971

Assagioli, a founder of transpersonal psychology, adds a vital dimension to humanistic psychology — the dimension of the metaphysical, the spiritual element at work in each person.

• Frankl, Charles. *The Case for Modern Man.* Beacon Press, Boston, 1955.

A deep search for a transcendent interpretation of human history, which places meaning as the key to human well-being.

IDEA

THERE ARE MULTITUDES OF SOLAR SYSTEMS. SOME MAY HAVE LIFE COMPARABLE TO OUR OWN.

In each of the billions of galaxies there are billions upon billions of solar systems — incalculable numbers of chances for other life. When we look up at the stars from our mother planet, we cannot see other intelligent beings. We couldn't even see life on Earth from the short distance of the moon!

We are able to measure the physical nature of stars. Through spectroscopes the white radiance of the cosmos can be shattered into rainbows, revealing by their spectral lines the identity of atoms pulsating billions of miles away. We are deciphering the story of starlight. The spectral lines inform us of the speed, temperature and composition of the stars. Most stars probably have planets with habitable zones where water can form, where gas can be held as atmosphere and radiation received in warming wave lengths.

In other words, the Universe is unitary on the physical level. There is no reason to assume that the bio-chemical processes which produced our bodies and minds are unique.

We are probably not alone. Whatever else lives here, in this Universe, is made of the same materials as our body; their atoms are our atoms; and their awareness is our awareness.

As we ourselves are born into the Universe physically, and begin gaining extraterrestrial capabilities, our expectancy of direct encounters with other life is intensifying. U.F.O. sightings increase. An estimated fifteen million Americans claim to have seen something inexplicable. Even NASA, our National Aeronautics and Space Administration, cannot discount all contacts as unreal.

The great extraterrestrial films — "2001 – A Space Oddyssey," "Startrek," "Star Wars," "Close Encounters" — attract millions of viewers.

Our cosmic sense is deepening as we ourselves are becoming young extraterrestrials. The vague premonition of beings from outer space is becoming more concrete to millions of people who are studying the skies at night for a sign. Many wish a U.F.O. would land in their front yard!

We are getting closer to contact with higher intelligence. We do not know what it is. There is a deep intuition in evolutionary people that "they" are like "us."

THIS WEEK

Remember any thoughts, signals, visions, sightings that have no explanation, especially those which had a deep impact on your life. Record them in the Journal and, if you like, share them with others in an Evolutionary Circle.

• • •

Imagine yourself at home. Suddenly you see vividly a strange light. "Something" unexplicable approaches and hovers above you. What is your reaction? What do you expect "them" to say to you? What do you say to them?

SUGGESTED READING:

• Sullivan, Walter. *We Are Not Alone.* McGraw-Hill Book Co. N.Y. 1964

• Hynek, Allen, J.. *The UFO Experience—A Scientific Inquiry.* Ballentine Books, N.Y. 1972

• Sagan, Carl, and Shklovskii. *Intelligent Life in the Universe.* Dell Publishing Co., Inc. N.Y. 1968

Each author is a serious, critical student of this fascinating subject. Several years ago no reputable scientist would dare study U.F.O.'s or extraterrestrials openly for fear of ridicule. Alan Hynek publishes the UFO Reporter and has the only U.F.O. Hotline in the world as fas as I know. He is computerizing all information to identify patterns. Also, he tries to get knowledgeable researchers to investigate the site. He is short of funds, but highly qualified and deserves major support. The U.F.O. Hotline number is (312)491-6666. The Sagan-Scklovskii book is important, because it is the first literary collaboration between American and Russian scientists on this subject. A major step forward in human relations as well as extraterrestrial!

IDEA

YOU ARE NOT ALONE, YOU ARE CONNECTED TO ALL BEINGS IN THE UNIVERSE AND MOST ESPECIALLY TO THOSE WHO DEEPLY ATTRACT YOU.

Empathy, "suprasexual attraction," is an evolutionary signal guiding us toward shared creative action with those who have a similar or related vocation. The future depends upon us making these connections. It is exceedingly difficult, if not impossible, to take your next step without deeper contact with those for whom you have genuine affinity.

The sense of aloneness, disconnectedness and depression is an evolutionary driver, urging you to take the initiative to reach out for those whom you need. They are there, needing you as much as you need them.

The need for connectedness is felt by all people. It is especially intense for those in whom the sense of the future is primary. The state of "evolutionary consciousness" is relatively new. Many of us have felt alone most of our lives. Existing institutions and cultural symbols affirm either the secular/scientific or the spiritual/religious. We sense both as intertwined. We are attempting to tune in to the evolving aspect of God, to become co-creative with the process of transformation. But there are as yet no social forms, no rituals to affirm our state of being. We are creating a first "School for Conscious Evolution," an invisible college for ACEs — agents of conscious evolution.

THIS WEEK

Think of your own attraction for the future. Let your mind wander through your whole life. Who do you know that shares this magnetic and "mysterious sense of the future." Imagine the most attractive, beautiful place you know. In your mind's eye invite all those people to join you in that place. Imagine sharing your experiences with each other in perfect spontaneity and trust. Feel the bonding that occurs in this imagined party. These connections are psychologically real and can be recalled at will, so that you never forget that you are not separate and alone. You are connected by deep bonds of affinity and mutual attraction to many people.

• • •

Start a list — a Book of Connections — of those people in real life whom you feel drawn to and sense you would like to know and work with. The list can include both those you know personally, and people you have heard of or read about, and would like to meet. Describe each person briefly — especially any significant encounter you may have had.

• • •

Try in whatever ways seem natural to contact them and share your thoughts with them. Trust the process to guide you. Your action will reinforce their own awareness and help meet a vital need in brother and sister co-evolvers — to overcome the illusion of aloneness and affirm evolutionary consciousness. Each person's Book of Connections can be shared with others, eventually circulated among all evolutionary journalists.

SUGGESTED READING:

• Teilhard de Chardin. *The Future of Man*. Harper Colophon Books, N.Y. 1969
Building the Earth. Avon, N.Y. 1969
The Phenomenon of Man. Harper and Row, N.Y. 1969

Teilhard de Chardin is the spiritual godfather of evolutionary futurism. A Catholic, Jesuit paleontologist whose work was not permitted to be published by the Church, he describes the evolution of the whole Universe as the struggle to become conscious partners with God. He combines science and religion in a way which forms the basis for evolutionary philosophy and politics. His *Building the Earth* is a call for spiritual action so persuasive that in an earlier presidential campaign one of the candidates, Sargent Shriver, used it as his text. It forms the guide for a "politics of hope." His master work is *The Phenomenon of Man*. It is required reading for all evolutionaries. *The Future of Man* is the best and easiest introduction to Teilhard's thought.

IDEA TO UNDERSTAND OURSELVES AS A WHOLE, WE STEP OUTSIDE OUR SYSTEM.

To see any body whole, you move beyond it. From the perspective within your body, you would not have the slightest idea what you look like as a whole. So, to understand ourselves as an interrelated planetary system, we stand outside ourselves in time/space and look upon ourselves as a whole body — Earth, a living organism of almost infinite diversity of which each being is an integral part.

THIS WEEK

Do a "planetary scan" each morning. Imagine yourself as a satellite orbiting the Earth, falling gracefully, freely without friction, weightless, as in fact our new sensing systems do, such as Landsat — monitoring weather, crop diseases, navigation, nuclear explosions, etc. Extend your own sensory system to "feel" your own planetary body intimately.

• • •

How do you feel today? How are we doing? Are we sick or well? Each of us is, organically, an interrelated member of this planetary body. Whatever is happening to the body, anywhere, is affecting each member. Can you experience yourself as part of this body?

• • •

When you listen to the "news" on radio or TV, realize it is a report on your larger body. Is it accurate? Are we getting the whole story — or is our mass media, our nervous system, still so infantile that it is picking up mostly pain, fear, and dissension, while ignoring our growth, creativity and successful efforts to overcome problems?

• • •

Identify the news story of this week that has most significance for our future. Write it as if for TIME Magazine — from the evolutionary perspective. Remember to view problems as evolutionary drivers, and new solutions as efforts to innovate and transform. If any of you are journalists by profession, see if this approach can find a place in your own news media.

SUGGESTED READING:

• Hubbard, Barbara Marx. *The Patterns of Creation – A History of the Future.* (To be published, contact Futures Network, 2325 Porter St., N.W., Washington D.C. 20008.)

A fresh look at the current world crises from the evolutionary perspective, focussing on evolutionary politics, psychology and spirituality, taken from an eight week course given at Georgetown University in 1981.

• Esfandiary, F.M. *Optimism One.* Popular Library, New York, 1970.

The first work of this revolutionary futurist who sees us transforming from the creature human condition to a Universal Humanity. The foundations for modern optimism.

IDEA

TAKE THE PERSPECTIVE OF A MORE MATURE SPECIES. WHAT MIGHT WE BE LIKE WHEN WE GROW UP, STOP FIGHTING AND START USING ALL OUR CAPACITIES HARMONIOUSLY?

In order to understand ourselves at this early phase of our universal history, take a moment to envision Humanity as we can become as we mature through this critical transition from Earth-only to universal life. What would it be like if everything we now know we can do, worked? This would include education, health, communications, productivity, systems planning, space, cybernetics, genetics, spirituality, psychic abilities, etc.

Our total capacities as a species include not only what we, four and a half billion individuals, now can do. It is also a composite gestalt — the combined capacities of cooperating peoples of the Solar System working together for our common development, with the awareness that we are part of the same creative process, born into the same Universe, with an indefinable future, physically and spiritually in a cosmos of incalculable abundance.

Probably no Earth-person has yet caught a clear glimpse of the wonder of a fully functioning planetary species at the next phase of evolution with all systems "go." A self revelation awaits us.

Thorton Wilder wrote "Our Town" from the perspective of the present looking at the past. We are taking the perspective of ourselves in the future — healthy, growing, each member alive and creative, looking back upon the present as our past. Knowing what we can do at our best, now, makes our present behavior actually archaic — the warring, the maldistribution of food, the traffic jams, etc. We are all masquerading as less than we are.

THIS WEEK

We assume that the future is not predetermined but prepatterned. The fifteen billion year process of transformation is continuing in our age. We have learned to expect synthesis, newness, a rise in consciousness, freedom and beauty at every step of the spiral. The process is prophetic and we are the product of the process. The story of creation is written in our blood and bones. In some sense, then, we already "know" the next step for our species, as a young child "knows" it will grow. Try to trigger your "memory of the future."

• • •

Experience yourself in the future. What is it like to have overcome hunger, disease, war, poverty? To live in space? To be able to extend your lifespan? To increase empathy and extrasensory capacities? To communicate with other life directly? To be liberated from involuntary work? To be attuning more directly to the evolutionary process? Let your deepest intuition guide you. Write down all your thoughts without censoring any.

• • •

Only science fiction and great religious visions of the future, such as the New Jerusalem in the Book of Revelations, describe this state of being. Religious visions of the future end where we are beginning — at the successful completion of this quantum transformation.

In our age, we are beginning to act out the intimations of transcendence — of a life beyond this life of eating, sleeping, reproducing and dying. Our new capacities in science and technology are providing us with the tools to act out the visions. Our expanding consciousness is providing us with the awareness we need. Be a cosmic anthropologist and write the outline of a book on Universal Humanity in the year 2025 — imagining that we made it through the quantum leap and everything is working well. (This takes genius, of course. But we've all got it!)

SUGGESTED READING:

• Lindsay, Hal. *There's A New World Coming — A Prophetic Odyssey*. Bantam Books, 1972.

The important sections in Lindsay's book are chapters 21 and 22. He outlines "God's Plan" as written in the Book of Revelation. Evolutionary thinkers should compare the Plan, especially the positive outcome of the holocaust — the New Jerusalem, where the Tree of Life grows, where there is no physical death, and where only the believers are present — with our projection of our planet at the next stage of evolution. It fits!

• Esfandiary, F.M. *Upwingers*. Popular Library, New York, 1977.

F.M. Esfandiary is a radical, brilliant evolutionary who projects the next step of evolution in concrete terms — what will happen with families, education, marriage, community and politics?

• Villoldo, Alberto, and Dychtwald, Ken. *Millenium*. J.P. Tarcher, Los Angeles, 1980.

An excellent compendium of "Glimpses into the 21st Century" by a host of evolutionary thinkers.

• Leary, Timothy. *Exo-Psychology*. Starseed/Peace Press, Calif., 1977.

Timothy Leary, after the trauma of the sixties, spent four years in prison. While in solitary confinement he broke through to an evolutionary vision of the future, which he calls "SMI²LE" — Space Migration, Intelligence Increase, Life Extension. Exo-Psychology. This is the first effort on Planet Earth to work out a psychology for the next stages of evolution, a milestone of thought, whether it is accurate or not.

IDEA

PERSONAL AND PLANETARY GROWTH ARE INSEPARABLE.

As each of us has a higher, wiser self toward which we are growing. Our species as a whole also has inherent growth potential. We are part of a planetary body undergoing the transition from Earth-only to universal life. Our own sense of inner transformation is literally an aspect of our planet's maturation. On the individual level, recent discoveries in humanistic and transpersonal psychologies reveal that each person has a natural need and ability to grow toward higher functioning and chosen creative work beyond meeting deficiency needs.

On a planetary scale, recent discoveries by general systems thinkers such as R. Buckminster Fuller, suggest that 100% of Humanity — our whole species — can live free from lack, a total success, by learning to cooperate with the laws, of the Universe.

Personally, and as a planet, we can go beyond meeting deficiency needs. "The ultimate creative capacity of the brain may be, for all practical purposes, unlimited," says George Leonard.

So are our physical capacities. Once we become an Earth-space people, using the abundant resources of a solar ecology rather than a closed, Earth-only ecology, we will have access to cosmic abundance which is the natural inheritance of a universal species.

Hundreds of "growth centers" for expanded wellness, personal enhancement, creativity, and spiritual attunement have sprung up in the past ten years, nurturing individuals toward a "high norm."

This phenomenon is occurring at the same time as our space scientists are establishing the ability to live in the expanded physical environment, our molecular biologists are unravelling the language of our genes, and cyberneticians are extending mental capacities.

From an evolutionary perspective, this coincidence of inner and outer growth is not an accident, but part of the same organic process of quantum transformation, part of the same cosmic birth.

The potential to become more, which each of us feels, sometimes as joy, and often as frustration, is, in collective form, the impulse of a species on the threshold of universal life.

THIS WEEK

Review your personal evolutionary spiral. See if you wish to change anything and especially if you can envision more clearly your own next steps? Has the "joy out front", that which is coming for you to do and become, clarified in any way? Can you sense the relationship between your own growing creativity and the physical expansion of Humanity's capabilities?

SUGGESTED READING:

- Fuller, R. Buckminster. *Utopia or Oblivion.* Bantam Books, N.Y. 1969

  It Came to Pass Not to Stay. MacMillan, N.Y. 1976

  Critical Path. St. Martin's Press, N.Y. 1981

"Bucky" Fuller is the closest person to a living evolutionary avatar on Planet Earth. An engineer and philosopher, he lays out the coming possibilities for our species as one hundred percent successful. He is the best doctor we have to guide us through the period of our birth into the Universe.

• O'Neill, Gerard K. *The High Frontier.* A Bantam Book, N.Y. 1974

Gerard O'Neill, a distinguished professor of physics at Princeton, dramatized the fact that space is a medium for growth and that a planetary surface may not be the proper place for the next phase of a technological society. Working with his students he developed the concept of space colonization using nonterrestrial resources, leading toward unlimited new worlds for the human race. His precise mind, his rapier intelligence, and his slight resemblance to Mr. Spock of Star Trek, has made him, much to his embarrassment, a popular hero — one of this age's great "overcomers" of self-imposed limits. If Gerry is right, the human race will never end.

IDEA AS A SPECIES, HUMANITY IS VERY YOUNG.

"The mind must stretch its concepts of space and time far beyond their normal limits to comprehend the sweep of the events that make up the history of our Universe. Suppose we adopt a point of view so broad that the tremendous span of a galaxy seems a detail, and the passage of a billion years is like an hour. Imagine the face of a cosmic clock on which one twenty-four hour day represents the life of the Universe. On this clock, one million years is a minute, and ten thousand years — the entire span of human civilization — is one tenth of a second."

> "Consider the great events in the history of life on the Earth within the framework of that analogy. Let the creation of the Universe occur at midnight; then the galaxies, stars and planets begin to form twenty minutes after midnight, and continue to form throughout the night and day. At four P.M. on the following afternoon, the sun, the Earth, and the moon appear. At 11:53 P.M. the fishes crawl out of the water; two minutes before midnight, the dinosaurs appear; sixty seconds later, they disappear; one second before midnight, modern man appears on the scene." (Jastrow, *Until the Sun Dies*)

Another way to look at the same facts is suggested by Peter Vajk's observation that our sun is five billion years old. Its life expectancy is another ten billion. In this time sequence it's 8:00 A.M.; the biosphere has awakened to self-awareness in Homo sapiens, and our work is about to begin. The main lesson is that we are at the beginning of our history as Humanity. Perhaps our imperfections and infantile behavior are the natural, awkward, dangerous dawn of a creature whose capacities are so enormous that until we become self-aware we will be self-destructive, like a newborn child.

THIS WEEK

List some obvious symptoms of our immaturity.

• • •

Where do you see signs of growing maturity?

SUGGESTED READING:

• Leakey, Richard E., and Lewin, Roger. *Origins — What New Discoveries Reveal About the Emergence of Our Species and Its Possible Future.* E.P. Dutton. N.Y. 1977

Richard Leakey's book will bring you up to date on the "mapping" of our origins. The information is still tantalizingly incomplete.

• Vajk, Peter J. *Doomsday has Been Cancelled.* Peace Press. CA. 1978

Peter Vajk is a courageous young space scientist who gave up a promising position with Lawrence Livermore Laboratory to dedicate himself to research in space development and related fields. His first book is a deep probing of how evolution works and is now working to provide opportunity for a universal future.

IDEA

OUR STORY OF CREATION COVERS BILLIONS OF YEARS. IT IS AN UNFINISHED STORY.

Most myths of creation have a beginning and an end:

> In the beginning God created the heaven and the earth . . .
> And on the seventh day God ended his work which he had made; and he rested on the seventh day from all his work which he had made.

From the evolutionary perspective our story of Creation has a beginning — but no known ending. We are in the midst of continuing Creation. This story of our genesis is a miracle play in which we are now conscious actors. Our every thought and act affects the future. There is no way we can get out of the play. We are learning the script of our past and co-authoring the script of the future.

Why is it that pre-scientific myths of creation had a beginning and an end, while ours appears to be continuing? Is it because, through expanding knowledge and power, we are awakening to the fact of our participation in designing the future?

THIS WEEK

Take the universal perspective and contemplate ourselves as co-creators of the future of Earth. How would you tell the story of this continuing creation to a child?

SUGGESTED READING:

- Campbell, Joseph. *Myths to Live By.* Bantam, New York. 1973

The work of a master teacher in the meaning of myths and symbols.

- Clarke, Arthur. *Childhood's End.* Ballantine Books, N.Y. 1972

One of the best visionary science fiction works on the evolutionary transformation of Homo sapiens to a new species.

IDEA

THE PURPOSE OF OUR POWER IS UNIVERSAL LIFE

All myths and early religions as well as the great world religions expressed intuitive knowledge of a life beyond this life, of reunion, of contract with higher forms of intelligence, and of transcendence of the human condition through ethical action, faith and mystery.

From 4,000 B.C. through 500 A.D., the Egyptians, Tibetans, Hindus, Hebrews, Christians and Buddhists translated the intimations and revelations of their great seers of cosmic life, who told of one Unitary Intelligence and purpose for Humanity, into theology, dogma and institutions. The translations, regardless of their inadequacies, nevertheless motivated billions of people throughout the Earth to aspire to something beyond their current and precarious condition. The intimations of a "next step," of an ethical "Last Judgment" on Earth-bound life, of expectations of immortality and of encounters with extraterrestrial and extrasensory fathers, mothers, angels, spirits, and gods runs consistently through human history — an irrepressible aspiration and experience.

With the advent of science and technology, these aspirations were discounted as unrealistic. Yet, we find that the very science and technologies which denied religious forecasts of the future are now providing us with those capacities we need to transcend our limitations — abilities heretofore relegated to "the gods." The technologists of transcendence are now studying the aging process to extend the life span, computerizing information, sending it in a "light beam" through laser technology, and ascending and descending alive from this Earth into the Universe in "chariots of flaming fire." They're sending images around the world and into the Universe with the speed of light, building three-dimensional light images — holograms — that look like "visions," learning how to reproduce asexually through cloning and other techniques, learning how to control our mental states, and even how to resurrect bodies from what used to be termed death by medical doctors. Scientists of the Universal Age are exploring the spectrum of human potential and fusing physical and psychic potentials to foresee a future for Humanity of universal consciousness and action.

THIS WEEK

How do you feel about the possibility of becoming a member of a universal species?

If the options became available, would you choose to:

Extend your life span?

Live in outer space, eventually, travelling beyond our solar system?

Communicate with other intelligent life?

Or would you prefer to remain on Earth and die?

This is a real question. Such choices may be available in our lifetime.

SUGGESTED READING:

• Wescott, Roger. *The Divine Animal: An Exploration of Human Potentiality*. Funk and Wagnalls Publishing Co., N.Y. 1969

Roger Wescott, who teaches at Drew University in Madison, New Jersey, is one of the great evolutionary anthropologists. His erudition is stunning. He can be the father of a new field of futuristic anthropology.

• Wilson, Robert Anton. *The Cosmic Trigger*. And/Or Press. Berkeley, Calif. 1977.

Robert Anton Wilson is another evolutionary pioneer. He probes the mysteries of extrasensory and extraterrestrial possibilities like a Sherlock Holmes on the ultimate mystery story! Very entertaining.

IDEA **THE PROCESS IS PROPHETIC.**

The billions of years of evolutionary transformations teach us how to predict coming transformations. The process is orderly. It obeys laws. Once the laws are understood, we can forecast the directions and tendencies of the process of change and work with them, as we now predict the behavior of objects in motion. In studying the evolutionary spiral, we have seen, over and over again, the same patterns reoccur: crisis . . . response through innovation . . . synthesis of separate elements into a more comprehensive whole . . . rise in consciousness and freedom.

We have witnessed the creation of new forms out of old.

We have seen that incremental changes do not lead up to more of the same.

Evolutionary studies provide the basis for a new futurism. It is rooted in knowledge of the whole past, human and pre-human, guided by spiritual resonance with the Designing Intelligence, informed by science. In can provide us with criteria with which to decide what is good and what may be destructive. Such guides are essential, for in the age of Conscious Evolution, we have become at least partially responsible for the future. Free choice increases along with power. Increasingly, the knowledge of the laws of evolutionary transformation will be regarded as vital for the survival of Humanity.

THIS WEEK

Examine some of the new capacities which are potentially highly destructive or creative. Remember to analyze them in relation to the criteria provided by the laws of evolutionary transformation.

- conception of children outside the womb
- life-extension
- space migration
- increase of psychic powers
- others — make your own list

What decisions would you make in terms of the research and development needed for these possibilities?

SUGGESTED READING:

• Land, George T. Lock. *Grow Or Die — The Unifying Principle of Transformation.* Dell Delta, New York (second edition), 1973.

George Land has been called the "Darwin" of our age. In his magnum opus, Grow Or Die, he identifies the pattern of growth in all entities from crystals to civilization and applies it to human problems in education, business, politics. This book is a must.

• Smuts, Jan. *Holism and Evolution.* Greenwood Press, Westport, Conn., 1973. (Reprint of 1926 edition).

Jan Smuts was, believe it or not, a famous leader in South Africa, as well as a relatively unknown thinker of the highest evolutionary significance. In this great work he identifies the underlying tendency of nature to form whole systems. A beautiful guide to our future.

IDEA

WHAT IN OUR AGE IS COMPARABLE TO THE BIRTH OF CHRIST? WHAT IS OUR STORY? IT IS A BIRTH, OUR OWN, THE BIRTH OF HUMANITY.

Christ brought the message that a time will come when we will know we are members of one body, we will forgive ourselves, and we will gain eternal life in partnership with God the Father. A transformation will occur in real time in real terms in history.

His message spread throughout the Earth for almost two thousand years. A flame of expectation arose of a "Second Coming" — the end of the world of separation and hate, the beginning of new worlds of love and eternal life. Only the timing was uncertain.

This hope lodged itself in the human heart and became the spiritual magnet for the Christian world. This life was seen as a preparation for the next. From an evolutionary perspective, the present quantum transformation is the time that has been awaited. It is the time of the Second Coming, the transition from Earth-bound to universal life.

At this moment of our planetary birth each person is called upon to recognize that the "Messiah is within." Christ-consciousness or cosmic consciousness is awakening in millions of Christians and non-Christians.

Through the new capacities being gained by science in the last few decades, we are developing the physical capacities for a life beyond this mammalian condition.

And because of the crises on our planet — nuclear, environmental, and social — we are being driven, pragmatically, to act out the vision of oneness and transcendence, or go to hell, literally, whether through nuclear holocaust, environmental collapse or social revolution.

In response, we are striving mightily to become more like Christ, to live in a co-creative relationship with God, building the New Jerusalem where "there shall be no more death, neither sorrows, nor crying, neither shall there be any more pain: for the former things are passed away." (Revelation. Chapter 21, verse 4.)

THIS WEEK

I have completed an evolutionary interpretation of the New Testament. It is a first future-oriented revisioning of the Judeo-Christian eschatology. It has transformed my life because I realize there is a model of the transformed human which I had been expecting. It is Jesus. This is no obscure model . . . obviously Christian-Western civilization has been built around it. It is a model of transformation of the person and the world. Writing it has accelerated my personal evolution. I now believe each of us is potentially a natural Christ.

• • •

Perhaps Evolutionary Journalists will contribute to this great task. It can only be written by those who combine spiritual and scientific awareness with an expectation of something to come. A first step is to list all the emerging capacities of the human race — including cosmic consciousness, action, time, community, contact, labor and individuality as identified in Chapter 5, "The Future: Previews of Coming Attractions." Begin to consider how these capacities can help fulfill the Judeo/Christian expectation, not as a metaphor, but as a real life transformation with everything working according to the "Plan" for planetary transformation at this stage of growth.

• • •

Relate your own personal sense of expectancy, the "joy out front," the magnet that attracts you, with the vision of the New Jerusalem built through conscious co-creatorship with the God-force. Does it resonate? Or is your expectation different?

SUGGESTED READING

• *The Bible.* "The Sermon on the Mount." Matthew, Chapters 5-7.

The greatest vision of the ethics of a co-creator, Humanity, at the next stage of moral evolution wherein we experience ourselves as one with God, each other and the whole Creation.

• Hubbard, Barbara Marx. *The Half Hour of Silence – An Evolutionary Interpretation of the New Testament.*

A new view of Jesus as our "potential self," a first vision of science and technology as an integral element in Humanity's actual transformation from mortal animal bodies to self-regenerating, universal beings in the model of Jesus. (To be published.)

TRAVEL GUIDE

Evolutionary Resources

THE EVOLUTIONARY PERSONALITY

As you proceed on your journey, let's take a look at you as a potential ACE, (Agent of Conscious Evolution,) as Barry McWater calls it. Let's see how you match up with the archetypal evolutionary personality.

Check the descriptions that fit you best:

☐ You relate the fulfillment of the potential of the world to the realization of your own capacities. You are not self-sacrificing but self-actualizing. Your moments of joy are when you forget your "self" and feel at one with the flow of the process.

☐ You have experienced deep empathetic communion with others. Your primary reward for action is the pleasure of loving, extending, deepening your creativity, empathy, connectedness.

☐ You have a deep respect for rational scientific inquiry and you also respect intuitive, holistic, psychic awareness.

☐ You are a risk-taker. You put yourself into unknown positions at the expense of security. You are willing to commit beyond your known abilities, risking failure.

☐ You are relatively well-balanced. You have learned to handle tensions that come into your life as you step over one abyss after another.

☐ You are struggling with the man-woman relationship in you personal life. You are not comfortable with the traditional male-female, dominant-submissive roles, nor are you satisfied with superficial relationships. You are aiming toward a deeper, non-passive, intimate union in which each partner is whole, each has synthesized the yin and yang, the feminine and masculine principles. You crave to unite whole being to whole being.

☐ You are relatively self-disciplined. You do not accept authoritarian discipline from outside. You do not accept a total master, but rather have many teachers. You find most of your wisdom in the collective wisdom of the social body, and within yourself.

☐ You don't feel completely contained within any existing institutional form of religion.

☐ You have a deep intuition of not quite belonging on Planet Earth, of being here at this time for a purpose beyond your full awareness.

☐ You have an expectation of newness. You do not conceive of existence as eternally repetitive.

☐ You are interested in transforming the world, not to reach a state of comfort for all but the opportunity for transcendence for all. You believe in the importance of material and social well-being, but you do not consider it the ultimate goal in life, nor do you believe people will be "satisfied" when their deficiency needs are well-met.

☐ You believe that the basis of reality is not material but consciousness and intelligence. You sense a spiritual reality.

☐ You believe you are not alone in the Universe. You intuitively sense the existence of other intelligence.

☐ You would accept an invitation from extraterrestrial visitors to go up in a space craft.

☐ If the option of an extended life span is available you would take it. You would choose extended physical life over death by disease. You would not be afraid to live in space, extend your life and seek out other life.

☐ You are fascinated with the teleology, or purpose, in religions, their "visions of the future," such as the New Jerusalem in the Book of Revelations.

☐ You identify with the traditional religious view that there is a God operating in the Universe and that there is a new state of being coming. You see the individual human and the collective human society playing a critical role in the transformation. We are free to abort the process or to nurture it to its natural next step.

☐ You relate the knowledge that has been gained through the last three hundred years of science, industry and technology to the religious visions of the future. You see science as intrinsic to transcending the limits of the animal human condition.

☐ You are physically healthy. You have a positive attitude toward life and practice some form of meditation or positive thinking exercises.

☐ You have a high sexuality. You are excited and exciting to others. But sexuality is not your primary goal.

☐ You have a long history of aloneness. You are an unusual type in your organization or family.

☐ You feel a deep need to stay in touch with others who share your attraction, because your potential cannot be released at the next level without interaction with others.

☐ You are highly individualistic, but cooperate easily. You don't have much of an "ego problem." You don't have to be dominant or submissive, but prefer working with equals or those who are desirous of becoming equals.

☐ You are not seeking manipulative power. You are interested in empowerment of yourself and others.

☐ You are not revolutionary but evolutionary. You are not interested in tearing down the system but working slightly beyond it and within it to evolve it with as much grace and as little trauma as possible.

☐ You have an intuition of a Designing Intelligence inherent in the nature of reality which is progressing toward a desirable future.

☐ You are attracted to know how that Intelligence works so that you can work with it toward a future which you sense as magnetic, a desirable state of being.

☐ You don't feel you are working alone in a universe with no pattern to guide you.

☐ You sense yourself on the verge of discovering something which is pre-figured and evolving, but which needs your energy to happen.

☐ You have fallen in love with possibilities and with the joy of participating with the process in achieving the next step.

☐ You don't know exactly what you are moving toward. The nature of the "goal" is open, not defined.

☐ You accept and even like to move toward the unknown. Not only can you tolerate ambiguity — you will create it whenever you find yourself settling into a fixed and static situation.

☐ You are willing to assume a lot of responsibility.

☐ You take initiative. You don't wait to be asked if you feel something needs to be done.

☐ You sense your responsibility not as a lonely burden, but as a partnership with the Creative Process, an Intelligence beyond your rational mind.

• • •

In checking your alignment with the "archetypal evolutionary" you may have identified some areas to focus on as your personal evolutionary journey unfolds. May you be blessed in your work.

ORGANIZATIONAL NETWORKING

The Institute for the Study of Conscious Evolution

The Institute for the Study of Conscious Evolution (ISCE) offers a context for those interested and/or working in the field of Conscious Evolution to connect, meet and work together. Based in San Francisco, ISCE has members throughout the country and the world.

Contributing members receive GAIA — ISCE's quarterly publication that reports on the progress of all Institute sponsored projects, and serves as a forum for discussion of key issues by leading thinkers and professionals in the field.

The purpose of GAIA is to help unify and strengthen the vision that so many people are trying to hold, and to make that vision a reality.

Members also receive discounts on publications, lectures, seminars and courses of study. At our annual conference on Conscious Evolution, members who are acting as "agents of conscious evolution" (ACE's) in diverse fields meet to discuss leading-edge work and to celebrate our collective participation in Conscious Evolution.

At present, member groups are forming in cities across the country — thus provides an opportunity for like-spirited people to meet and collaborate.

You are invited to become a member of ISCE. In addition to providing a context for evolutionary networking, your membership contribution supports important evolutionary research and educational projects in personal transformation, intuitive knowing, man/woman relationships, organizational development, holistic governance and planetary ecology.

Simply send $25 or more (by check or major credit card #) to:

ISCE
2418 Clement Street
San Francisco, CA 94121

We thank you.

AN EVOLUTIONARY LIBRARY

THE BIBLE, The Book of Revelation. (King James Version.) An early vision of the "New Jerusalem," the next step of evolution.

Blair, Lawrence, *RHYTHMS OF VISION*. Warner Books, New York, 1975. A beautifully written description of the interrelated design of nature.

Bohm, David, *WHOLENESS AND THE IMPLICATE ORDER*. Rutledge and Kegan, Boston, 1980. A quantum physicist's fascinating theory of a metaphysical substrate to the material world (which sounds surprisingly like God or consciousness.)

Bovra, Ben. *THE HIGH ROAD*. Houghton Mifflin Co., Boston, 1981. Excellent rationale for space as an answer to social problems.

Bucke, R.M., *COSMIC CONSCIOUSNESS*. Dutton, New York, 1969. A seminal work on the evolution of consciousness with a focus on key individuals who have achieved cosmic consciousness. Essential reading.

Campbell, Joseph, *MYTHS TO LIVE BY*. Bantam, New York, 1973. The work of a master-teacher in the meaning of human myths and symbols.

Capra, Fritjof, *THE TAO OF PHYSICS*. Bantam, New York, 1977. A brilliant description of the relationsip between modern science and mysticism.

Clarke, Arthur, C., *CHILDHOOD'S END.* Ballantine Books, New York, 1972. One of the best visionary science-fiction works on the evolutionary transformation of Homo sapiens to a new species.

Coulter, Arthur N., *SYNERGETICS.* Prentice Hall, New Jersey, 1976. Practical methods to achieve deep group cooperation.

THE COURSE IN MIRACLES. Foundation for Inner Peace, Tiburon, California, 1975. A monumental three volume work chanelled by a Columbia University psychologist teaching forgiveness, oneness, Christ-consciousness through deep inner listening for the voice of God within. A workbook consisting of 365 lessons, a text of the philosophy, and a manual for teachers.

de Chardin, Teilhard, *BUILDING THE EARTH.* Avon, New York, 1969. Teilhard's approach to social transformation through the self-actualization of Humanity.

de Chardin, Teilhard, *THE APPEARANCE OF MAN.* Harper, New York, 1967. Anthropology from the evolutionary spiritual perspective.

de Chardin, Teilhard, *THE FUTURE OF MAN.* Harper and Row, New York, 1969. A good introduction to Teilhard's ideas.

de Chardin, Teilhard, *THE PHENOMENON OF MAN.* Harper and Row, New York, 1969. The seminal work of one of the key modern thinkers in evolutionary spirituality. His magnum opus. Essential reading.

Esfandiary, F.M., *OPTIMISM ONE.* Popular Library, New York, 1978. Esfandiary's major ideas.

Esfandiary, F.M., *TELESPHERES.* Popular Library, New York, 1977. A further elaboration of *Upwingers.*

Esfandiary, F.M., *UPWINGERS.* Popular Library, 1977. A radical vision of Humanity as a universal, immortal species with pragmatic ideas as to social, personal, educational and behaviorial changes. Essential reading.

Fabel, Arthur, *COSMIC GENESIS: Teilhard de Chardin and the Emerging Scientific Paradigm.* The American Teilhard Association for the Future of Man, Inc. by ANIMA Books, Chambersburg, Pennsylvania, 1980. An excellent bibliography on Conscious Evolution.

Ferguson, Marilyn, *THE AQUARIAN CONSPIRACY.* J.P. Tarcher, Los Angeles, California, 1980. The best book on the human potential aspect of the transformation including networking possibilities. Essential reading.

Fuller, R. Buckminster, *AND IT CAME TO PASS — NOT TO STAY.* Macmillan Publishing Co., Inc., New York, 1976. A beautiful poetic essay encapsulating Bucky's ideas.

Fuller, R. Buckminster, *THE CRITICAL PATH.* St. Martin's Press, New York, 1981. Bucky's life's thought and work, well organized and presented. Essential reading.

Fuller, R. Buckminster, *UTOPIA OR OBLIVION.* Bantam, New York, 1969. A good shorter version of Bucky's basic thought.

Glenn, Jerome Clayton, *SPACE TREK.* Stackpole, Pennsylvania, 1978. An excellent description of the benefits of space development from the economic, political and defense perspective.

Gorney, Roderic, *THE HUMAN AGENDA.* Bantam, New York, 1973. A master work depicting human potential in the context of cosmic evolution.

Harrington, Alan, *THE IMMORALIST.* Random House, New York, 1969. A brilliant description of the desirability, the pragmatic feasibility and the naturalness of human life extension.

Houston, Jean, *LIFEFORCE.* Delacorte, New York, 1980. An excellent work which gives the reader an understanding of human potential and exercises to develop it.

Houston, Jean and Masters, Robert, *MIND GAMES.* Dell Publishing Co., Inc., New York, 1973. A wonderful introduction to "how to" expand your own mind potential.

Hubbard, Barbara, *THE HUNGER OF EVE.* Stackpole, Pennsylvania, 1976. One of the first personal descriptions of the transformation of a traditional woman's life as wife, mother and helper to an evolutionary cosmic futurist.

Hubbard, Earl, *THE CREATIVE INTENTION.* Interbook, New York, 1976. An artist-philosopher's poetic statement of human evolution and the need for "new worlds" in space.

Hubbard, Earl, *OUR NEED FOR NEW WORLDS.* Interbook, New York, 1976. A more pragmatic description of the same theme as The Creative Intention.

Hynek, J. Alan, *THE UFO EXPERIENCE.* Ballentine Books, New York, 1972. A basic work on the UFO question by the world's leading scientific authority.

Jantsch, Erich, *CONSCIOUSNESS AND EVOLUTION.* (Edited by Jantsch and C.H. Waddington), Addison-Wesley Publishing Co., Massachusetts, 1976. An excellent anthology of key scientific, mathematical views on how evolution works.

Jantsch, Erich, *DESIGN FOR EVOLUTION.* Braziller, New York, 1976. A brilliant presentation of evolutionary philosophy from the general systems perspective.

Jastrow, Robert, *GOD AND THE ASTRONOMERS.* Warner, 1980. This brilliant astronomer peers beneath the manifesting world to seek the Creator.

Jastrow, Robert, *UNTIL THE SUN DIES.* W. W. Norton, New York, 1977. A poetic scientific essay on cosmology and the future by a leading NASA scientist and philosopher.

Koestler, Arthur, *JANUS.* Random House, New York, 1978. Essays into the wholistic nature of reality.

Land, George T. Lock, *GROW OR DIE.* Dell Delta, New York, (second edition), 1973. One of the most comprehensive synthesis ever attempted, identifying recurring patterns of growth, applied to current problems. Essential reading.

Leakey, Richard and Lewin, Roger, *ORIGINS.* E. P. Dutton, New York, 1977. An excellent introduction to the latest scientific information on human origins.

Leary, Timothy, *EXO-PSYCHOLOGY.* A Starseed/Peace Publication, Culver City, California, 1977. A Western master's version of transformation from Homo sapiens to Homo universalis through the evolution of the human nervous system. A presentation of the next stage of evolution.

SMI^{2}LE: (Space Migration, Intelligence Increase, Life Extension.) Must reading.

Linsey, Hal, *THERE'S A NEW WORLD COMING.* Bantam, New York, 1975. The Christian fundamentalist view of the transformation.

Lipnack, Jessica, and Stamps, Jeffrey. *NETWORKING.* A Dolphin Book, Doubleday & Co., Inc. Garden City, New York, 1982. A brilliant guide to joining the networks of shared attraction — from space to environment, art, education, health, etc.

Lovelock James, *GAIA.* Oxford, 1977. A fascinating theory that the Earth has the characteristics of a living organism.

Maslow, Abraham H., *RELIGIONS, VALUES AND PEAK EXPERIENCES.* Viking Press, New York, 1970. The spiritual aspects of human psychology.

Maslow, Abraham H., *TOWARD A PSYCHOLOGY OF BEING.* D. Van Nostrand Co., New York, 1971. The seminal work that launched the "third force" in psychology: humanistic and transpersonal psychology. Essential reading.

McWaters, Barry, *CONSCIOUS EVOLUTION.* Evolutionary Press, San Francisco, 1982. A pioneering work by the founder of a new school of thought which sees the individual as an evolving participant in the evolution of the whole world. Combines esoteric, environmental, psychological and social insights in a brilliant synthesis. Essential reading.

Needleman, Jacob, *A SENSE OF THE COSMOS.* Doubleday and Co., New York, 1975. An interesting approach to the synthesis of scientific and spiritual insights.

O'Neill, Gerard K., *THE HIGH FRONTIER.* Bantam, New York, 1977. A good description of the purpose and potential of space colonies.

Rosenfeld, Alfred, *THE SECOND GENESIS.* Pyramid Communications, New York, 1972. A basic work on the impact of the biological revolution on our lives by one of the world's best science writers.

Rosenfeld, Alfred, *PROLONGEVITY*. Knopf, New York, 1975. An excellent overview of recent research in the field of aging and life extension.

Sauber, William, *THE FOURTH KINGDOM*. Aquari Corp., Box 1966, Midland, Michigan, 1975. An original and provocative insight that technology is a natural "kingdom" like the mineral, vegetable and animal, and that its primary purpose is to seed Earth-life in the Universe via space arks.

Sheldrake, Rupert, *A NEW SCIENCE OF LIFE*. Blond & Briggs, London, 1981. The important theory of "morphic resonance fields."

Smuts, Jan, *HOLISM AND EVOLUTION*. Greenwood Press, Connecticut, 1973, (reprint of 1926 edition.) A profoundly significant work on nature's intrinsic tendency to evolve through the formation of whole systems.

Stine, G. Harry, *THE THIRD INDUSTRIAL REVOLUTION*. G.P. Putnam's Sons, New York, 1975. The basic reader in the potential of space industrialization.

Sullivan, Walter, *WE ARE NOT ALONE*. McGraw Hill, New York, 1964. Excellent summary of the possibility of extraterrestrial life by a New York Times science writer.

Toffler, Alvin, *THE THIRD WAVE*. William Morrow & Co., New York, 1980. A classic in the field of social transformation.

Vajk, J. Peter, *DOOMSDAY HAS BEEN CANCELLED*. Peace Press, Culver City, CA, 1977. A good description of the hope for a universal, unlimited future as a natural evolutionary step through appropriate development of social and scientific tools including space colonization. Overcomes the space/Earth arguement.

Villoldo, Alberto, and Dychtwald, Ken, eds., *MILLENIUM.* J.P. Tarcher, Inc., Los Angeles, 1980. A good anthology of transformational thinkers.

Westcott, Roger, *THE DIVINE ANIMAL.* Funk and Wagnalls, New York, 1969. A brilliant scholarly look at human development, viewing us as Earth's ambassadors to the Universe.

Whyte, Lancelot Law, *THE NEXT DEVELOPMENT IN MAN.* A Mentor Book, New American Library, Inc., New York, 1945. A profound analysis of the unitary nature of nature and history. Essential reading.

Wilson, Robert Anton, *THE COSMIC TRIGGER.* And/Or Press, Berkeley, CA, 1977. A fascinating group of essays on esoteric, exoteric, occult and scientific possibilities.

Wilson, Robert Anton, *THE ILLUMINATTI PAPERS.* And/Or Press, Berkely, CA, 1980. Another fascinating presentation of esoteric evolutionary ideas.

Young, Arthur, *THE REFLEXIVE UNIVERSE.* Delacorte Press, San Francisco, 1977. A technological genius looks at the fundamental nature of evolution.

Zukav, Gary, *THE DANCING WU LI MASTERS: An Overview of the New Physics*, Bantam Books, Inc., New York, 1979. A wonderfully clear overview of the new physics. Essential reading.

Key Publications

BRAIN/MIND BULLETIN. Interface Press, P.O. Box 42211, Los Angeles, CA 90042

FUTURE LIFE. 475 Park Ave., South, New York, NY 10016

THE FUTURIST. World Future Society, P.O. Box 30369, Washington, D.C. 20014

GAIA. Institute for the Study of Conscious Evolution, 2418 Clement St., San Francisco, CA 94121

INSTITUTE OF NOETIC SCIENCES NEWSLETTER. 2824 Union St., San Francisco, CA 94123

JOURNAL OF THE BRITISH INTERPLANETARY SOCIETY. 27/29 South Lambeth Rd., London, 5W8 1SZ, England

L-5 NEWS. L-5 Society, 1060 East Elm, Tucson, AZ 85719

LEADING EDGE. Interface Press, P.O. Box 42211, Los Angeles, CA 90042

NEW SCIENTIST. 19 Oxford St., London, WCIA 1NG, England

NEXT. P.O. Box 10045, Des Moines, Iowa 50340

OMNI. Omni Publications International, Ltd., 909 Third Ave., New York, NY 10022

QUEST/82. P.O. Box 2448, Boulder, CO 80302

REVISION. P.O. Box 316, Cambridge, MA 02138

TECHNOLOGY REVIEW, MIT. Cambridge, MA 02139